SpringerBriefs in Energy

SpringerBriefs in Energy presents concise summaries of cutting-edge research and practical applications in all aspects of Energy. Featuring compact volumes of 50 to 125 pages, the series covers a range of content from professional to academic. Typical topics might include:

- A snapshot of a hot or emerging topic
- A contextual literature review
- A timely report of state-of-the art analytical techniques
- An in-depth case study
- A presentation of core concepts that students must understand in order to make independent contributions.

Briefs allow authors to present their ideas and readers to absorb them with minimal time investment.

Briefs will be published as part of Springer's eBook collection, with millions of users worldwide. In addition, Briefs will be available for individual print and electronic purchase. Briefs are characterized by fast, global electronic dissemination, standard publishing contracts, easy-to-use manuscript preparation and formatting guidelines, and expedited production schedules. We aim for publication 8–12 weeks after acceptance.

Both solicited and unsolicited manuscripts are considered for publication in this series. Briefs can also arise from the scale up of a planned chapter. Instead of simply contributing to an edited volume, the author gets an authored book with the space necessary to provide more data, fundamentals and background on the subject, methodology, future outlook, etc.

SpringerBriefs in Energy contains a distinct subseries focusing on Energy Analysis and edited by Charles Hall, State University of New York. Books for this subseries will emphasize quantitative accounting of energy use and availability, including the potential and limitations of new technologies in terms of energy returned on energy invested. The second distinct subseries connected to SpringerBriefs in Energy, entitled Computational Modeling of Energy Systems, is edited by Thomas Nagel, and Haibing Shao, Helmholtz Centre for Environmental Research— UFZ, Leipzig, Germany. This sub-series publishes titles focusing on the role that computer-aided engineering (CAE) plays in advancing various engineering sectors, particularly in the context of transforming energy systems towards renewable sources, decentralized landscapes, and smart grids.

All Springer brief titles should undergo standard single-blind peer-review to ensure high scientific quality by at least two experts in the field.

Xiaoyu Wang · Yufei Wang · Qi Liang ·
Yuning Zhang

Fundamentals of Single Cavitation Bubble Dynamics

 Springer

Xiaoyu Wang
Key Laboratory of Power Station Energy
Transfer Conversion and System (Ministry
of Education)
School of Energy Power and Mechanical
Engineering
North China Electric Power University
Beijing, China

Qi Liang
Key Laboratory of Power Station Energy
Transfer Conversion and System (Ministry
of Education)
School of Energy Power and Mechanical
Engineering
North China Electric Power University
Beijing, China

Yufei Wang
Key Laboratory of Power Station Energy
Transfer Conversion and System (Ministry
of Education)
School of Energy Power and Mechanical
Engineering
North China Electric Power University
Beijing, China

Yuning Zhang ⓘD
Key Laboratory of Power Station Energy
Transfer Conversion and System (Ministry
of Education)
School of Energy Power and Mechanical
Engineering
North China Electric Power University
Beijing, China

ISSN 2191-5520   ISSN 2191-5539 (electronic)
SpringerBriefs in Energy
ISBN 978-3-031-75040-3   ISBN 978-3-031-75041-0 (eBook)
https://doi.org/10.1007/978-3-031-75041-0

© The Editor(s) (if applicable) and The Author(s), under exclusive license to Springer Nature
Switzerland AG 2024

This work is subject to copyright. All rights are solely and exclusively licensed by the Publisher, whether
the whole or part of the material is concerned, specifically the rights of translation, reprinting, reuse
of illustrations, recitation, broadcasting, reproduction on microfilms or in any other physical way, and
transmission or information storage and retrieval, electronic adaptation, computer software, or by similar
or dissimilar methodology now known or hereafter developed.
The use of general descriptive names, registered names, trademarks, service marks, etc. in this publication
does not imply, even in the absence of a specific statement, that such names are exempt from the relevant
protective laws and regulations and therefore free for general use.
The publisher, the authors and the editors are safe to assume that the advice and information in this book
are believed to be true and accurate at the date of publication. Neither the publisher nor the authors or
the editors give a warranty, expressed or implied, with respect to the material contained herein or for any
errors or omissions that may have been made. The publisher remains neutral with regard to jurisdictional
claims in published maps and institutional affiliations.

This Springer imprint is published by the registered company Springer Nature Switzerland AG
The registered company address is: Gewerbestrasse 11, 6330 Cham, Switzerland

If disposing of this product, please recycle the paper.

Preface

Cavitation and bubble dynamics have important scientific significance and application value in the fields of fluid machinery and fuel atomization. This book provides a comprehensive review of the rapidly expanding field of single cavitation bubble dynamics, covering the discussion of equations of bubble dynamics, bubble oscillation dynamics, theoretical prediction models of jets, and high-speed photography technology. Among them, the core formulas, important research methods, and typical results related to bubble oscillation and collapse dynamics are systematically and comprehensively introduced. Specifically, in terms of the equations of bubble dynamics, several classical dynamic equations utilized to describe the radial motion of the spherical bubble, cylindrical bubble, and the bubble in a droplet are derived and compared. In terms of the bubble oscillation dynamics, based on the perturbation method, multi-scale method, and Laplace transform method, the nonlinear oscillation characteristics of the bubble in free oscillation and driven oscillation are analyzed respectively. In terms of the jet prediction theory, the Kelvin impulse model and various boundary treatment methods are given in detail, and the jet direction, intensity, and spatial sensitivity caused by the bubble collapse near various boundaries are discussed. In terms of the bubble collapse visualization based on high-speed photography, taking the laser-induced bubble technology as an example, the system composition, operation process, and experimental layout of the high-speed photography experimental platform are introduced, and a large number of typical bubble collapse deformation, jet evolution, and shock wave propagation characteristics obtained from experiments are demonstrated. This book is intended for academic researchers and graduate students in fluid dynamics, aiming to consolidate the basic theory, physical mechanism, and latest progress in the field of bubble dynamics.

Beijing, China
September 2024

Xiaoyu Wang
Yufei Wang
Qi Liang
Yuning Zhang

Acknowledgement This book was financially supported by the National Natural Science Foundation of China (Project No.: 51976056).

Contents

About the Authors

Xiaoyu Wang received his doctor's degree from North China Electric Power University in June 2024. He focuses his research interests on cavitation and bubble dynamics, concentrating on experimental research, theoretical analysis, and numerical simulation of high-speed phenomena such as bubble oscillation, jets, and shock waves. He has participated in the publication of 17 journal papers in these fields, bubble resonance characteristics, which discovered the jet deflection phenomena and the formation mechanism of the shock waves.

Yufei Wang is currently a master degree candidate at North China Electric Power University. She mainly engages in research on the cavitation bubble morphology evolution near various types of walls and bubble collapse dynamics. She has participated in the publication of two journal papers in these fields, which investigated the bubble morphology evolution and the non-spherical behaviors of a cavitation bubble.

Qi Liang is currently a doctoral candidate at North China Electric Power University. She mainly focuses on research on the cavitation bubble collapse dynamics and the collapse jet characteristics. She has published three journal papers in these fields, investigating the strength and directional deflection properties of jet near the particle and the wall.

Yuning Zhang is currently a professor at North China Electric Power University. He primarily focuses on the research in cavitation and bubble dynamics. He has published two monographs in Springer Press and over 90 papers (ten highly cited papers and one hot paper) in journals such as *Nature Communications*, *Physics of Fluids*, and *Energy*. He was ranked by Stanford University as one of the "top 2% of scientists in the world by 2023." In addition, he is an associate editor of *IET Renewable Power Generation* and an editorial board member of six international/national journals such as the *Journal of Hydrodynamics*. He also serves as the chairman of the International Cavitation Forum 2016, a member of the organizing committee of the WIMRC International Symposium of Cavitation (UK) and many other international conferences, and has given six invited talks on international conferences. He has received many awards from the Society of Energy and Electric Power in China.

Chapter 1
Introduction

1.1 Research Background

When the local pressure in a liquid is lower than the saturated vapor pressure at that temperature or when external high energy enters, the fluid starts to vaporize and forms cavitation bubbles, which is called the cavitation phenomenon [1]. In engineering production, the cavitation phenomena are often accompanied by liquid flow and have a wide range of applications.

Figure 1.1 shows the extensive applications of the cavitation phenomena in different fields. In the hydraulic machinery field, core components of the equipment such as pump impellers [2, 3], propellers [4], and hydrofoils [5–7] are often exposed to cavitation phenomena due to their high-speed rotation or displacement movement. The cavitation bubbles rapidly collapse under pressure changes and produce high-speed jets, causing a continuous impact on the component surface. Under long-term impact, metal surfaces may gradually erode, forming pits and cracks, which not only reduces the service life of the equipment, but can also lead to malfunctions and safety issues. In the biomedical field, the energy of the bubble collapse is applied to help dissolve blood clots for ultrasonic thrombolysis procedures [8] as well as to crush kidney stones [9, 10]. These non-invasive treatment options can reduce pain and risk for patients. In addition, cell rupture technology promotes the release of intracellular substances through the cavitation effects, which has potential applications in drug delivery and gene therapy [11]. In the surface cleaning field, cavitation bubbles can improve cleaning efficiency and reduce the use of chemical cleaning agents [12–15]. This method can effectively remove dirt and bacteria from the surface while reducing the impact on the environment. In the material processing field, cavitation bubbles can be applied to processes such as drilling, slotting, and cutting [16]. Compared with traditional machining, it can provide higher precision and less material damage. In the alloy strengthening field, cavitation bubbles can form tiny holes and cracks on the alloy surface, thus enhancing the hardness and wear resistance of materials [17, 18]. In the sonochemistry field, the creation and rupture of cavitation bubbles can

© The Author(s), under exclusive license to Springer Nature Switzerland AG 2024
X. Wang et al., *Fundamentals of Single Cavitation Bubble Dynamics*,
SpringerBriefs in Energy, https://doi.org/10.1007/978-3-031-75041-0_1

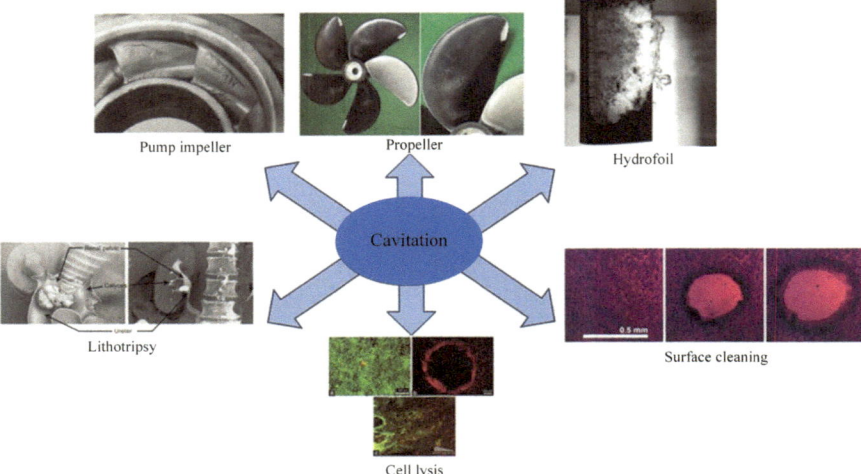

Pump impeller

Propeller

Hydrofoil

Cavitation

Lithotripsy

Surface cleaning

Cell lysis

Fig. 1.1 Cavitation phenomena in different fields. Reprinted with the permission from Ref. [22] Copyright (2024) (Springer Singapore)

facilitate chemical reactions [19], which are widely used in synthetic chemistry and environmental treatment. In the shipbuilding industry field, proper cavitation can help to improve the cleanliness of ship hulls and reduce biological contamination [20, 21], which can improve navigational efficiency and reduce the maintenance cost of ships. In summary, the dynamics of the cavitation bubble show its unique advantages and broad application prospects in many fields.

1.2 Research Status

For the cavitation phenomenon, the oscillation and collapse dynamics of a single cavitation bubble are of central concern. As listed in Table 1.1, in this research field, a series of important contributions have been achieved, with a focus on properties of cavitation nucleation, jets, shock waves, and bubble morphological evolution. In terms of cavitation nucleation, Li et al. [23] found that the presence of particles has an important effect on the cavitation nucleation process and the nucleation rate will increase. In terms of jets, typical types include regular jets, needle jets, and counter jets, which are significantly affected by the wall geometry and the relative position of the bubble to the wall. Specifically, Wang et al. [24] found a needle jet generated before a regular jet using femtosecond illumination. It can cause a severe water hammer pressure of more than 100 MPa and dissipate rapidly within 800 ns, which is supposed to be the direct cause of the erosion marks on the material surface. Zhang [25] observed the counter jet appearing in the bubble rebound stage and found that the maximum impact pressure on the wall could reach 1 GPa. In terms of shock

Table 1.1 Current status of research on cavitation bubble dynamics

Phenomenon	Type	References
Cavitation nucleation	Nucleation sites	[23, 30–32]
	Nucleation rates	
	Nucleation thresholds	
Jets	Regular jets	[24, 33, 33–39]
	Needle jets	
	Counter jets	
Shock waves	Shock waves during bubble inception	[26–28, 40–44]
	Shock waves during bubble collapse	
Morphological evolution	Neck structure	[45, 29]
	Bubble separation	
	Particle rupture	

waves, they will be excited during the bubble growth and collapse stages. Specifically, Yu et al. [26] described the formation mechanism of multiple shock waves during the bubble collapse, including the process of jet formation, jet piercing the bubble surface, and jet impacting the wall. Lai et al. [27] observed the rapid evolution of shock waves in time and space during the expansion stage of a cavitation bubble and found that the normalized shock wave width is not related to the laser energy and duration. Zou et al. [28] found that the presence of rigid boundaries altered the consistency of the liquid medium around the bubble, resulting in the emission of water hammer and implosion shock waves when the bubble breaks, respectively. This stratification effect is closely related to the bubble-boundary distance and the boundary structure. In terms of bubble morphological evolution, the cavitation bubble will collapse in a complex non-spherical shape when it is near a boundary. Zheng et al. [29] investigated the process of bubble collapse between two particles and found that the bubble can present different collapse shapes as spherical, olive, drum, and gyroscopic, respectively, due to the different distances between the bubble and the particles.

1.3 Description of the Book

The main chapters of the book are organized as follows: this chapter discusses the practical applications of cavitation and bubble dynamics in engineering, briefly reviews the research status of single cavity dynamics, and gives a brief description of the chapter structure of this book. Chapter 2 derives the radial equations of motion from different perspectives for different types of bubbles including the spherical bubble, the cylindrical bubble, and the bubble within a droplet. Chapter 3 introduces several approximate solution methods for equations of bubble dynamics and analyzes the free and driven oscillation characteristics of the bubbles. Chapter 4 presents a jet prediction method based on the Kelvin impulse theory and the mathematical treatment of different boundaries. In addition, taking several typical boundaries as examples, the distribution characteristics of the jet direction and intensity are analyzed. Chapter 5 introduces the core equipment and operating principles of the high-speed photography experimental system, and shows the characteristics of the cavitation bubble collapse deformation, jet evolution, shock wave propagation, and droplet splashing. Chapter 6 sums up the main conclusions of the book.

References

1. Brennen CE (2014) Cavitation and bubble dynamics. Cambridge University Press, Cambridge
2. Binama M, Muhirwa A, Bisengimana E (2016) Cavitation effects in centrifugal pumps: a review. Int J Eng Res Appl 6(5):52–63
3. Fu X, Li D, Wang H et al (2021) Hydraulic fluctuations during the pump power-off runaway transient process of a pump turbine with consideration of cavitation effects. J Hydrodyn 33(6):1162–1175
4. Pfitsch W, Gowing S, Fry D et al (2009) Development of measurement techniques for studying propeller erosion damage in severe wake fields. In: CAV2009-7th international symposium on cavitation, Michigan
5. Venning JA, Pearce BW, Brandner PA (2022) Nucleation effects on cloud cavitation about a hydrofoil. J Fluid Mech 947:A1
6. Yin T, Pavesi G, Pei J et al (2022) Large eddy simulation of cloud cavitation and wake vortex cavitation around a trailing-truncated hydrofoil. J Hydrodyn 34(5):893–903
7. Xie N, Tang Y, Liu Y (2023) High-fidelity numerical simulation of unsteady cavitating flow around a hydrofoil. J Hydrodyn 35(1):1–16
8. Suo D, Jin Z, Jiang X et al (2017) Microbubble mediated dual-frequency high intensity focused ultrasound thrombolysis: an in vitro study. Appl Phys Lett 110(2):023703
9. Sagar H, Moctar O (2023) Dynamics of a cavitation bubble between oblique plates. Phys Fluids 35(1):1–28
10. Loske AM (2010) The role of energy density and acoustic cavitation in shock wave lithotripsy. Ultrasonics 50(2):300–305
11. Jasikova D, Rysová M, Kotek M (2019) Application of laser-induced breakdown cavitation bubbles for cell lysis in vitro. Int J Appl Pharm 11(5):186–190
12. Lohse D, Janve V, Arora M et al (2006) Surface cleaning from laser-induced cavitation bubbles. Appl Phys Lett 89(7):074102
13. Verhaagen B, Rivas DF (2016) Measuring cavitation and its cleaning effect. Ultrason Sonochem 29:619–628
14. Cui P, Zhang A, Wang S et al (2018) Ice breaking by a collapsing bubble. J Fluid Mech 841:287–309
15. Song WD, Hong MH, Lukyanchuk B et al (2004) Laser-induced cavitation bubbles for cleaning of solid surfaces. J Appl Phys 95(6):2952–2956
16. Charee W, Tangwarodomnukun V (2018) Dynamic features of bubble induced by a nanosecond pulse laser in still and flowing water. Opt Laser Technol 100:230–243
17. Ren X, Wang J, Yuan S et al (2018) Mechanical effect of laser-induced cavitation bubble of 2A02 alloy. Opt Laser Technol 105:180–184
18. Xu W, Tzanakis I, Srirangam P et al (2016) Synchrotron quantification of ultrasound cavitation and bubble dynamics in Al–10Cu melts. Ultrason Sonochem 31:355–361

19. Sutkar VS, Gogate PR (2010) Mapping of cavitational activity in high frequency sonochemical reactor. Chem Eng J 158(2):296–304
20. Tian Z, Liu Y, Zhang A et al (2020) Jet development and impact load of underwater explosion bubble on solid wall. Appl Ocean Res 95:102013
21. Zhang A, Zeng L, Cheng X et al (2011) The evaluation method of total damage to ship in underwater explosion. Appl Ocean Res 33(4):240–251
22. Yu J, Wang X, Hu J et al (2023) Laser-induced cavitation bubble near boundaries. J Hydrodyn 35(5):858–875
23. Li B, Gu Y, Chen M (2019) Cavitation inception of water with solid nanoparticles: a molecular dynamics study. Ultrason Sonochem 51:120–128
24. Wang J, Wang G, Zeng Q et al (2023) Recent progress on the jetting of single deformed cavitation bubbles near boundaries. J Hydrodyn 35(5):832–857
25. Zhang J (2021) Effect of stand-off distance on "counterjet" and high impact pressure by a numerical study of laser-induced cavitation bubble near a wall. Int J Multiphase Flow 142:103706
26. Yu J, Hu J, Liu Y et al (2024) Numerical investigations of the interactions between bubble induced shock waves and particle based on OpenFOAM. J Hydrodyn 36(2):355–362
27. Lai G, Geng S, Zheng H et al (2022) Early dynamics of a laser-induced underwater shock wave. J Fluids Eng 144(1):011501
28. Zou L, Luo J, Xu W et al (2023) Experimental study on influence of particle shape on shockwave from collapse of cavitation bubble. Ultrason Sonochem 101:106693
29. Zheng X, Wang X, Ding Z et al (2023) Investigation on the cavitation bubble collapse and the movement characteristics near spherical particles based on Weiss theorem. Ultrason Sonochem 93:106302
30. Borkent BM, Arora M, Ohl CD et al (2008) The acceleration of solid particles subjected to cavitation nucleation. Phys Fluids 610:157–182
31. Borkent BM, Arora M, Ohl CD (2007) Reproducible cavitation activity in water-particle suspensions. J Acoust Soc Am 121(3):1406–1412
32. Gu Y, Li B, Chen M (2016) An experimental study on the cavitation of water with effects of SiO_2 nanoparticles. Exp Therm Fluid Sci 79:195–201
33. Lauterborn W, Bolle H (1975) Experimental investigations of cavitation-bubble collapse in the neighbourhood of a solid boundary. J Fluid Mech 72(2):391–399
34. Dular M, Požar T, Zevnik J et al (2019) High-speed observation of damage created by a collapse of a single cavitation bubble. Wear 418–419:13–23
35. Reuter F, Deiter C, Ohl CD (2022) Cavitation erosion by shockwave self-focusing of a single bubble. Ultrason Sonochem 90:106131
36. Reuter F, Ohl CD (2021) Supersonic needle-jet generation with single cavitation bubbles. Appl Phys Lett 118(13):41542
37. Bußmann A, Riahi F, Gökce B et al (2023) Investigation of cavitation bubble dynamics near a solid wall by high-resolution numerical simulation. Phys Fluids 35(1):016115
38. Lindau O, Lauterborn W (2003) Cinematographic observation of the collapse and rebound of a laser-produced cavitation bubble near a wall. J Fluid Mech 479:327–348
39. Vogel A, Lauterborn W, Timm R (1989) Optical and acoustic investigations of the dynamics of laser-produced cavitation bubbles near a solid boundary. J Fluid Mech 206:299–338
40. Geng S, Yao Z, Zhong Q et al (2021) Propagation of shock wave at the cavitation bubble expansion stage induced by a nanosecond laser pulse. J Fluids Eng 143(5):051209
41. Ohl CD, Kurz T, Geisler R et al (1999) Bubble dynamics, shock waves and sonoluminescence. Philos Trans R Soc Lond A 357(1751):269–294
42. Yang X, Liu C, Wan D et al (2021) Numerical study of the shock wave and pressure induced by single bubble collapse near planar solid wall. Phys Fluids 33(7):073311
43. Yang X, Liu C, Li J et al (2023) Numerical study of liquid jet and shock wave induced by two-bubble collapse in open field. Int J Multiphase Flow 168:104584

44. Hu J, Lu X, Liu Y et al (2023) Numerical and experimental investigations on the jet and shock wave dynamics during the cavitation bubble collapsing near spherical particles based on OpenFOAM. Ultrason Sonochem 99:106576

45. Zhang Y, Chen F, Zhang Y et al (2018) Experimental investigations of interactions be-tween a laser-induced cavitation bubble and a spherical particle. Exp Therm Fluid Sci 98:645–661

Chapter 2
Equations of Bubble Dynamics

This chapter focuses on the dynamics equations of the bubbles with different types. Firstly, the Rayleigh-Plesset equation for a spherical bubble is constructed and the effects of the liquid compressibility on the equation results are further considered. On this basis, the differences between the equations with different precisions are analyzed. Secondly, three typical cylindrical equations of bubble dynamics are derived respectively. The differences and connections between these equations in form are discussed. Finally, the dynamics equations of the bubble inside a spherical droplet are established in detail.

2.1 Spherical Bubble

2.1.1 Rayleigh-Plesset Equation

In this section, a bubble is assumed to exist within an infinite liquid environment whose radius varies with time and can be expressed as "$R(t)$". The temperature and pressure are "T_∞" and "$p_\infty(t)$" respectively, where the temperature is constant and the pressure is a controlled and known input to control the growth process of the bubble. In addition, the liquid density is "ρ" and the dynamic viscosity of the liquid is "μ". The distance from the radial position of a point in the liquid to the bubble center is represented by "r". The pressure, outward radial velocity and temperature at a point inside the liquid can be expressed as "$p(r, t)$", "$u(r, t)$" and "$T(r, t)$" respectively. According to the law of conservation of mass, the local radial velocity of the liquid can be obtained as [1]

$$u(r, t) = \frac{F(t)}{r^2} \tag{2.1}$$

© The Author(s), under exclusive license to Springer Nature Switzerland AG 2024
X. Wang et al., *Fundamentals of Single Cavitation Bubble Dynamics*,
SpringerBriefs in Energy, https://doi.org/10.1007/978-3-031-75041-0_2

When the mass transfer process is ignored, the relationship between "$F(t)$" and "R" can be obtained as [1]

$$F(t) = R^2 \dot{R} \tag{2.2}$$

where "$\dot{R}$" and "$\ddot{R}$" represent the first-order and second-order derivatives to "R", respectively.

The internal gas growth during the expansion of the bubble is equal to the volume growth "$4\pi R^2 \dot{R}$". Therefore, the mass evaporation inside the bubble can be expressed as "$\rho_v(T_B) 4\pi R^2 \dot{R}$", where "$\rho_v(T_B)$" is the saturated vapor density at the temperature "T_B". The mass evaporation can be obtained as the mass flow rate of liquid through the interface into the bubble. The mass evaporation can be obtained as equal to the liquid mass flow rate into the bubble through the interface. Therefore, the inward velocity of the liquid relative to the interfacial interface is expressed as "$\dot{R}\rho_v(T_B)/\rho$", so we can obtain that [1]

$$u(R, t) = \dot{R} - \frac{\rho_V(T_B)}{\rho}\dot{R} = \left[1 - \frac{\rho_V(T_B)}{\rho}\right]\dot{R} \tag{2.3}$$

$$F(t) = \left[1 - \frac{\rho_v(T_B)}{\rho}\right]R^2\dot{R} \tag{2.4}$$

It always satisfies "$\rho_V(T_B) \ll \rho$" in the real case, so the approximate form of Eq. (2.2) has good generality.

The N–S equation for motion in the r-direction is [1]

$$-\frac{1}{\rho}\frac{\partial p}{\partial r} = \frac{\partial u}{\partial t} + u\frac{\partial u}{\partial r} - v\left[\frac{1}{r^2}\frac{\partial}{\partial r}\left(r^2\frac{\partial u}{\partial r}\right) - \frac{2u}{r^2}\right] \tag{2.5}$$

By substituting "$u = F(t)/r^2$" into Eq. (2.5), we can obtain that [1]

$$-\frac{1}{\rho}\frac{\partial p}{\partial r} = \frac{1}{r^2}\frac{dF}{dt} - \frac{2F^2}{r^5} \tag{2.6}$$

By integrating Eq. (2.6) we have [1]

$$\frac{p - p_\infty}{\rho} = \frac{1}{r}\frac{dF}{dt} - \frac{1}{2}\frac{F^2}{r^4} \tag{2.7}$$

Next, the dynamic boundary condition for the bubble surface is constructed. Assume a control body consists of a small, infinitely thin layer containing a portion of the interface. The outward radial force per unit area is given by [1]

$$(\sigma_{rr})_{r=R} + p_B - \frac{2S}{R} \tag{2.8}$$

Owing to "$\sigma_{rr} = -p + 2\mu \partial u / \partial r$", the combined force per unit area can also be written as [1]

$$p_B - (p)_{r=R} - \frac{4\mu}{R}\dot{R} - \frac{2S}{R} \tag{2.9}$$

When mass transport (evaporation or condensation) through the interface is not considered, this combined force of Eq. (2.9) is zero, while substituting "$(p)_{r=R} = p_B - \frac{4\mu}{R}\dot{R} - \frac{2S}{R}$" and "$F = R^2\dot{R}$" in Eq. (2.7), the generalized Rayleigh-Plesset equation can expressed as [1]

$$R\ddot{R} + \frac{3}{2}(\dot{R})^2 = \frac{p_\infty(t) - p_B(t)}{\rho} + \frac{4v}{R}\dot{R} + \frac{2S}{\rho R} \tag{2.10}$$

2.1.2 Equations Considering Liquid Compressibility

The following basic assumptions are used: (1) The bubble remains spherical during growth and collapse, with no movement of the bubble center. (2) The liquid is compressible. (3) Both mass transfer and thermal effects at the bubble–liquid interface are neglected. (4) The gas pressure is uniform within the bubble.

A spherical coordinate system is established with the center of the bubble as the origin, and the continuity equation and momentum equation of the liquid flow around the bubble are described as [2]

$$\frac{\partial \rho}{\partial t} + \frac{1}{r^2}\frac{\partial}{\partial r}\left(r^2 \rho u\right) = 0 \tag{2.11}$$

$$\frac{\partial u}{\partial t} + u\frac{\partial u}{\partial r} + \frac{1}{\rho}\frac{\partial p}{\partial r} = 0. \tag{2.12}$$

Here, the local radial velocity u can be expressed by the liquid velocity potential as "$u = \partial \varphi / \partial r$". In addition, for the thermodynamic properties of a liquid with reversible adiabatic process satisfy [2]

$$c^2 = \frac{dp}{d\rho} \tag{2.13}$$

$$dh = \frac{1}{\rho}dp \tag{2.14}$$

where "c" represents the local velocity of sound in the liquid; "h" represents the local specific enthalpy. In order to close the model, the following equation of state for water in Tate form is introduced as [3]

$$\frac{p+B}{p_\infty + B} = \left(\frac{\rho}{\rho_\infty}\right)^n, \tag{2.15}$$

$$p_\infty = p_0 + \rho_\infty g H. \tag{2.16}$$

Here, "p_0" is the atmospheric pressure; "g" is the local gravity acceleration; "H" is the underwater depth of the bubble. "B" and n are the empirical constants and satisfy $B = 3049 \times 10^5$ Pa and $n = 7.15$ for the water [4].

The gas pressure in the bubble can be expressed as [2]

$$p_i = \left(p_\infty + \frac{2\sigma}{R_0}\right)\left(\frac{R_0}{R}\right)^{3\kappa}. \tag{2.17}$$

where, "p_i" is the gas pressure in the bubble, "σ" represents the surface tension coefficient; "R_0" is the equilibrium instantaneous radius of the bubble; "κ" is the polytropic exponent, which satisfies $\kappa = 1.4$ during the adiabatic process.

The radial velocity of the liquid at the interface between the bubble and the liquid is equal to the velocity of the bubble wall movement, and the liquid pressure "p_B" can be expressed as [2]

$$p_B = p_i - \frac{2\sigma}{R} - \frac{4\mu\dot{R}}{R} \tag{2.18}$$

In order to facilitate the subsequent derivation, the following dimensionless parameters are introduced as [4]

$$r_* = \frac{r}{R_0}, \tag{2.19}$$

$$R_* = \frac{R}{R_0}, \tag{2.20}$$

$$t_* = \frac{t}{R_0/U}, \tag{2.21}$$

$$\varphi_* = \frac{\varphi}{R_0 U}, \tag{2.22}$$

$$h_* = \frac{h}{U^2}, \tag{2.23}$$

$$c_* = \frac{c}{c_\infty}, \tag{2.24}$$

$$\varepsilon = \frac{U}{c_\infty}, \tag{2.25}$$

$$u = \sqrt{\frac{p_\infty}{\rho_\infty}}. \tag{2.26}$$

Here, "ε" is the Mach number. "c_∞" represents the sound velocity in the liquid at infinity.

Combining Eqs. (2.13)–(2.15) and (2.23)–(2.26), the dimensionless specific enthalpy of the liquid at a bubble wall can be obtained as [2]

$$h_{B*} = \frac{1}{\varepsilon^2} \frac{1}{n-1} \left[\left(\frac{p_B + B}{p_\infty + B} \right)^{\frac{n-1}{n}} - 1 \right]. \tag{2.27}$$

Combining Eqs. (2.11)–(2.15), (2.19) and Eqs. (2.21)–(2.25), the dimensionless differential control equations for the liquid motion are expressed as [2]

$$\left(1 - \varepsilon^2(n-1) \left(\frac{\partial \varphi_*}{\partial t_*} + \frac{1}{2} \left(\frac{\partial \varphi_*}{\partial r_*} \right)^2 \right) \right) \nabla_*^2 \varphi_* = \varepsilon^2 \left(\frac{\partial^2 \varphi_*}{\partial t_*^2} \right.$$

$$\left. + 2 \frac{\partial \varphi_*}{\partial r_*} \frac{\partial^2 \varphi_*}{\partial r_* \partial t_*} + \left(\frac{\partial \varphi_*}{\partial r_*} \right)^2 \frac{\partial^2 \varphi_*}{\partial r_*^2} \right) \tag{2.28}$$

$$\frac{\partial \varphi_*}{\partial t_*} + \frac{1}{2} \left(\frac{\partial \varphi_*}{\partial r_*} \right)^2 + h_* = 0. \tag{2.29}$$

Next, the two singularly disturbed differential equations are simplified by employing the matched asymptotic approximation method. According to the order of the spatial variable "r_*", the fluid outside the bubble can be divided into two parts: an inner field near the bubble and an outer field far from the bubble. The approximate solutions exist in different forms because of the different approximation results and boundary conditions of the equations in the two fields. According to the perturbation theory, the solutions for the velocity potential and enthalpy of the inner field can be given as [4]

$$\varphi_* = \sum_{i=0}^{m} \varepsilon^i \varphi_i, \tag{2.30}$$

$$h_* = \sum_{i=0}^{m} \varepsilon^i h_i. \tag{2.31}$$

and for the external field, [4]

$$\varphi_* = \sum_{i=0}^{m} \varepsilon^i \phi_i, \tag{2.32}$$

$$h_* = \sum_{i=0}^{m} \varepsilon^i H_i. \tag{2.33}$$

Here, "m" is a natural number which controls the solution accuracy. Based on the matched asymptotic expansions method, the inner boundary of the outer field is equal to the outer boundary of the inner field [4]. When $m \leq 1$, the specific relationship is [4]

$$\lim_{r_* \to \infty} \frac{\sum_{i=0}^{m} \varepsilon^i \varphi_i}{\varepsilon^m} = \lim_{\varepsilon r_* \to 0} \frac{\sum_{i=0}^{m} \varepsilon^i \phi_i}{\varepsilon^m}, \tag{2.34}$$

$$\lim_{r_* \to \infty} \frac{\sum_{i=0}^{m} \varepsilon^i h_i}{\varepsilon^m} = \lim_{\varepsilon r_* \to 0} \frac{\sum_{i=0}^{m} \varepsilon^i H_i}{\varepsilon^m}. \tag{2.35}$$

when $m = 2$, [4]

$$\lim_{r_* \to \infty} \frac{\sum_{i=0}^{m} \varepsilon^i \varphi_i}{\varepsilon^m} = \lim_{\varepsilon r_* \to 0} \frac{\sum_{i=0}^{m-1} \varepsilon^i \phi_i}{\varepsilon^m} + \lim_{\varepsilon^{1/2} r_* \to 0} \phi_m, \tag{2.36}$$

$$\lim_{r_* \to \infty} \frac{\sum_{i=0}^{m} \varepsilon^i h_i}{\varepsilon^m} = \lim_{\varepsilon r_* \to 0} \frac{\sum_{i=0}^{m-1} \varepsilon^i H_i}{\varepsilon^m} + \lim_{\varepsilon^{1/2} r_* \to 0} H_m. \tag{2.37}$$

When taking $m = 0$ in Eqs. (2.30)–(2.33), only retained the terms containing "ε^0" in the equations. Combining Eqs. (2.28)–(2.35), we can derive solutions for the dimensionless velocity potential "φ_0" and the enthalpy "h_0" of the internal field, which has an accuracy of order 0 (simply called zero-order approximate solution). Further, by substituting the boundary condition equation Eq. (2.27) into the expression of "h_0", the dimensionless equations for the bubble wall motion in the case of free oscillation with zero-order accuracy can obtained as [5, 6]

$$R_* \ddot{R}_* + \frac{3}{2} \dot{R}_*^2 = h_{B*}. \tag{2.38}$$

When taking $m = 1$ in Eqs. (2.30)–(2.33), only the terms containing "ε^0" and "ε^1" are retained. In the internal field, the solutions for the dimensionless velocity potential "φ_1" and enthalpy "h_1" can be solved by Eqs. (2.28)–(2.35). Then, by substituting the boundary condition equation into the expression for "$h_0 + \varepsilon h_1$", the dimensionless bubble wall equations of motion with first-order accuracy for the bubble (short for one-order equations) can be obtained as [7]

$$\left[1 - (1 + \lambda)\varepsilon \dot{R}_* \right] R_* \ddot{R}_* + \frac{3}{2} \left[1 - \left(\frac{1}{3} + \lambda \right) \varepsilon \dot{R}_* \right] \dot{R}_*^2$$
$$= \left[1 + (1 - \lambda)\varepsilon \dot{R}_* \right] h_{B*} + \varepsilon R_* \dot{h}_{B*} \tag{2.39}$$

Here, "λ" is an arbitrary parameter with order less than "$1/\varepsilon$". Equation (2.39) is a set of first-order dimensionless bubble wall equations of motion. When taking $\lambda = 0$ and 1, Eq. (2.39) becomes the Keller-Miksis equation [8] and Herring equation [9] expressed in terms of enthalpy, respectively.

When taking $m = 2$ in Eqs. (2.30)–(2.33), the bubble wall equations of motion with second-order accuracy for freely oscillating bubble (short for second-order equations) can be obtained as [4]

$$
\left[1 - (1+\lambda)\varepsilon\dot{R}_* + \left(\frac{14}{5} + 2\lambda + \theta\right)\varepsilon^2\dot{R}_*^2 \right] R_*\ddot{R}_*
$$
$$
+ \frac{3}{2}\left[1 - \left(\frac{1}{3}+\lambda\right)\varepsilon\dot{R}_* + \left(\frac{16}{15} + \frac{4}{3}\lambda + \theta\right)\varepsilon^2\dot{R}_*^2 \right] \dot{R}_*^2
$$
$$
+ \varepsilon^2 R_*^2\ddot{R}_*^2 = \left[1 + (1-\lambda)\varepsilon\dot{R}_* + \theta\varepsilon^2\dot{R}_*^2 \right] h_{B*}
$$
$$
+ \varepsilon R_*\left[1 - (1+\lambda)\varepsilon\dot{R}_{**} \right]\dot{h}_{B*}. \tag{2.40}
$$

Here, "θ" is an arbitrary parameter with order less than "$1/\varepsilon$". When taking $\lambda = 1$ and $\theta = 0$, Eq. (2.40) is the bubble wall equation of motion in the Tomita form [10] and Fujikawa form [11].

2.1.3 Comparison of Equation Results

This section exhibits the difference between first-order and second-order equations by comparing the difference in their predicted values such as the radius of the bubble etc. Figure 2.1 shows the variation of the dimensionless bubble radius with dimensionless time. The solid and dashed lines in the figure correspond to the predictions of the first and second-order equations, respectively. Considering that the bubble radius does not vary much during the first period [4], the figure shows the data from the end of the first period. As can be seen from the figure, there is a significant difference between the predictions of the bubble radius by the two equations, which reflects the fact that their predictions of the dissipated power and resonance frequency are also significantly different.

Figure 2.2 shows the dimensionless radial velocity of the bubble wall with respect to the dimensionless time. The solid and dashed lines in the figure represent the predictions of the first and second-order equations, respectively. "$\dot{R}_{\mathrm{max,\,col*}}$" represents the maximum collapse velocity of the bubble wall. "$\dot{R}_{\mathrm{max,\,reb*}}$" represents the maximum rebound velocity of the bubble wall. As shown, the predictions of the two equations for the maximum collapse velocity, the maximum rebound velocity, and the length of the oscillation period of the bubble show significant differences. Through simulation analysis, it is found that the dimensionless difference in the maximum collapse velocity predicted by the two equations reaches 20.62%, while the dimensionless difference in the maximum rebound velocity is as high as 38.55%. With the

Fig. 2.1 The variation of the
dimensionless bubble radius
with dimensionless time.
Reprinted with the
permission from Ref. [2]
Copyright (2022) (Elsevier)

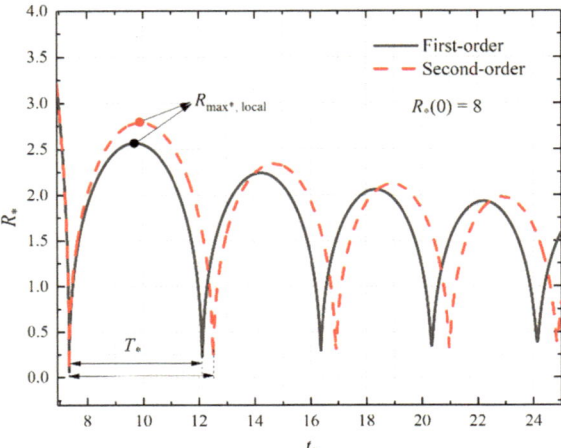

bubble oscillation continues, the energy of the bubble gradually decreases, leading
to a gradual decrease in the difference between the predictions of the first-order and
second-order equations.

Figure 2.3 shows the dimensionless radial acceleration of the bubble wall with
respect to dimensionless time. The solid and dashed lines in the figure correspond to
the predictions of the first-order and second-order equations, respectively. As shown
in Fig. 2.3, the radial acceleration predicted by the second-order equation is significantly lower than that of the first-order equation, and the dimensionless difference
between the two reaches 129.46%. This indicates that the difference between the
two equations in predicting the dimensionless radial acceleration is more significant
compared to the difference in predicting the dimensionless radial velocity of the
bubble radius and bubble wall.

Fig. 2.2 The dimensionless
radial velocity of the bubble
wall with respect to the
dimensionless time.
Reprinted with the
permission from Ref. [2]
Copyright (2022) (Elsevier)

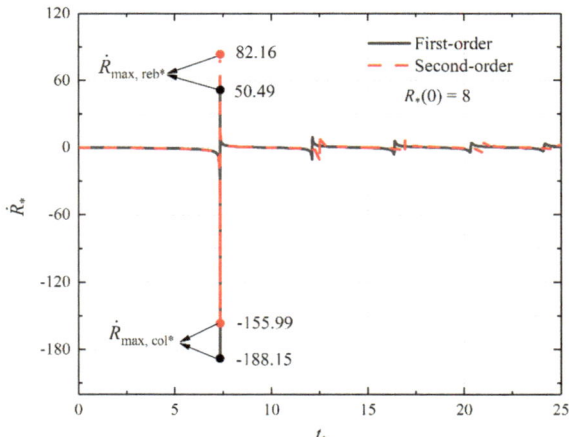

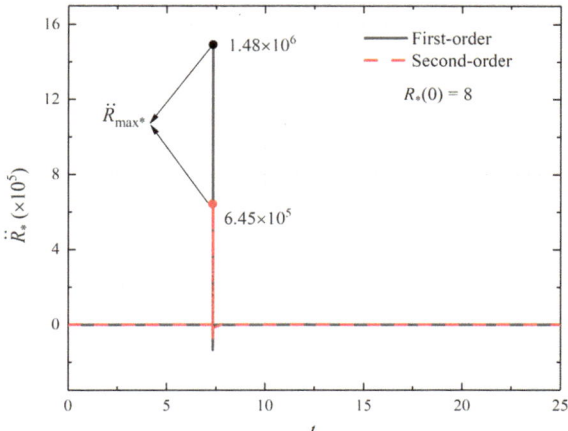

Fig. 2.3 The dimensionless radial acceleration of the bubble wall with respect to dimensionless time. Reprinted with the permission from Ref. [2] Copyright (2022) (Elsevier)

Figure 2.4 shows the trend of the maximum bubble wall Mach number with respect to the initial dimensionless bubble radius. The black solid line and the red dashed line in the figure represent the predictions of the first-order and second-order equations, respectively. The blue solid line represents the critical case where the maximum bubble wall Mach number reaches 1. It can be observed from the figure that the maximum bubble wall Mach number predicted by both equations rises as the bubble radius increases. Meanwhile, for the same bubble radius, the maximum bubble wall Mach number predicted by the first-order equation always exceeds that predicted by the second-order equation. And, the difference between the two increases gradually as the bubble radius increases. In addition, when the maximum bubble wall Mach number is equal to 1, the first-order equation corresponds to a dimensionless bubble radius of 6.9, whereas the second-order equation corresponds to 7.7. This means that the effective prediction range of the second-order equation is much wider than that of the first-order equation, depending on the principle that the maximum bubble wall Mach number should be less than 1 [12].

Figure 2.5 predicts the variation of dissipated power with initial dimensionless bubble radius for the first-order and second-order equations. The black solid line and the red dashed line in the figure represent the predictions of the first-order and second-order equations, respectively. Subgraph (a)–(d) refer to the thermal dissipated power E_{th}, the viscous dissipated power E_v, the radiation dissipated power E_r and the total dissipated power E_{tot}, respectively. The difference between the radiative dissipation power predicted by the two equations, as well as the total dissipation power, increases gradually as the bubble radius increases. And the difference in thermal dissipation power first increases as the bubble radius increases, but then decreases because the thermal dissipation power of the first-order equation decreases. As for the viscous dissipation power, there is almost no significant difference between the predictions of the two equations.

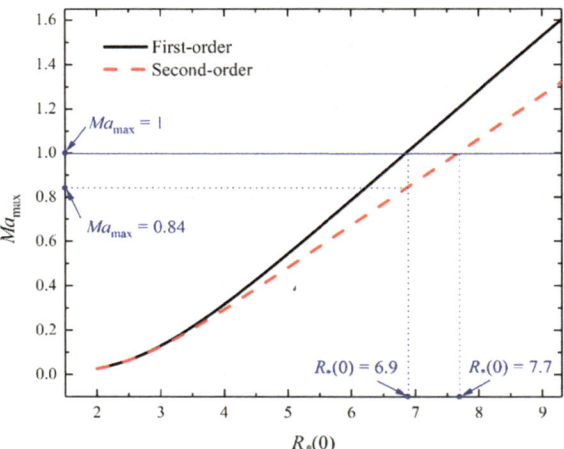

Fig. 2.4 The trend of the maximum bubble wall Mach number with respect to the initial dimensionless bubble radius. Reprinted with the permission from Ref. [2] Copyright (2022) (Elsevier)

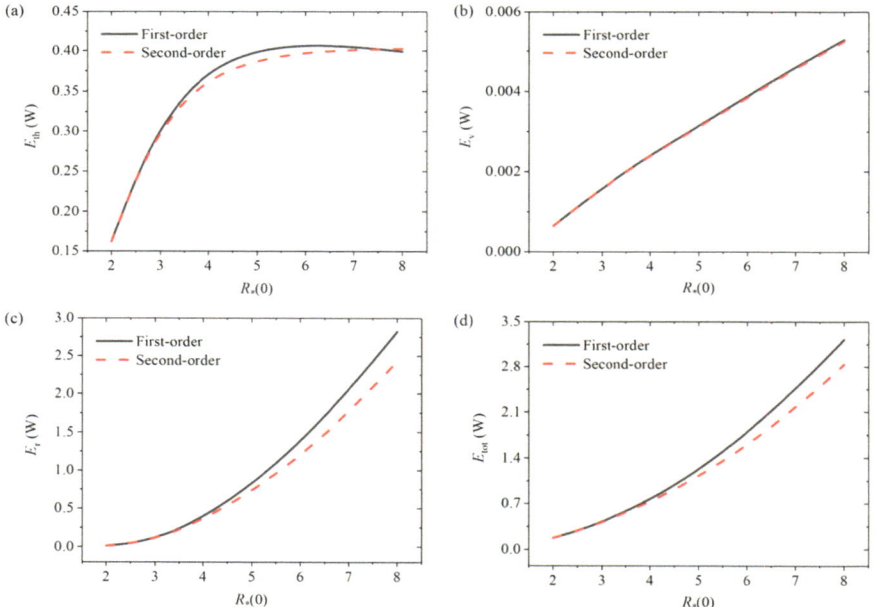

Fig. 2.5 The variation of dissipated power with initial dimensionless bubble radius. Reprinted with the permission from Ref. [2] Copyright (2022) (Elsevier)

2.2 Cylindrical Bubble

2.2.1 Equation from Momentum Perspective

The continuity and momentum equations can be expressed respectively as [13]

$$\frac{\partial \rho}{\partial t} + \rho \frac{\partial u}{\partial r} + u \frac{\partial \rho}{\partial r} + \frac{\rho u}{r} = 0 \tag{2.41}$$

$$\frac{\partial u}{\partial t} + u \frac{\partial u}{\partial r} + \frac{1}{\rho} \frac{\partial p}{\partial r} = 0 \tag{2.42}$$

When the fluid is assumed to be incompressible, Eq. (2.41) can be simplified as Eq. (2.43) [13]. And the boundary conditions for motion can be written as Eqs. (2.44) and (2.45) [6].

$$\frac{\partial u}{\partial r} + \frac{u}{r} = 0 \tag{2.43}$$

$$u(R, t) = \dot{R} \tag{2.44}$$

$$\frac{du(R, t)}{dt} = \ddot{R} \tag{2.45}$$

Multiplying "r" on both sides of Eq. (2.43) at the same time, a new expression can be obtained as [6]

$$\frac{\partial (ru)}{\partial r} = 0 \tag{2.46}$$

$$r \cdot u(r, t) = F(t) \tag{2.47}$$

By substituting Eq. (2.44) into Eq. (2.47), it gives [6]

$$u(r, t) = \frac{R\dot{R}}{r} \tag{2.48}$$

Substituting Eq. (2.48) into Eq. (2.42) and then integrating we can have [6]

$$\int_{R}^{r_0} \left(\frac{1}{r} (R\ddot{R} + \dot{R}^2) - \frac{1}{r^3} R^2 \dot{R}^2 \right) = -\frac{1}{\rho_0} \int_{p_{\text{ext}}}^{p_0} \frac{\partial p}{\partial r} \tag{2.49}$$

The equation of motion can be expressed as [6]

$$\left(R\ddot{R} + \dot{R}^2\right)\ln\frac{r_0}{R} - \frac{1}{2}\dot{R}^2 + \frac{R^2}{2r_0^2}\dot{R}^2 = \frac{p_{\text{ext}} - p_0}{\rho_0} \tag{2.50}$$

$$p_{\text{ext}} = p_{\text{in}} - \frac{\sigma}{R} - \frac{2\mu}{R}\dot{R} \tag{2.51}$$

$$p_{\text{in}} = \left(p_0 + \frac{\sigma}{R_0}\right)\left(\frac{R_0}{R}\right)^{2\kappa} \tag{2.52}$$

where "r_0" represents an artificially selected upper limit of integration, "p_0" represents the pressure at r_0, "κ" is the multivariate index with 1.4, "p_{in}" and "p_{ext}" represents the internal and external pressure of the bubble, respectively.

2.2.2 Equation from Energy Perspective

The energy perspective equation for a cylindrical bubble can be derived from the Lagrange equation, where the Lagrange equation is given as [13]

$$\frac{d\left(\partial L/\partial\dot{R}\right)}{dt} - \frac{\partial L}{\partial R} = 0 \tag{2.53}$$

where "$L = K - U$", "K" and "U" are kinetic energy and potential energy, respectively. When the bubble is placed between two parallel plates at a distance "D", the kinetic energy of the fluid is [14]

$$K = \frac{1}{2}\rho_0 D \int_R^\infty u^2 2\pi r\, dr = \pi\rho_0 DR^2\dot{R}^2 \int_R^\infty \frac{dr}{r} \tag{2.54}$$

When the upper limit of integration of Eq. (2.54) is set to a finite value, Eq. (2.54) can be simplified as [14]

$$K = \pi\rho_0 DR^2\dot{R}^2 \ln\frac{r_0}{R} \tag{2.55}$$

The bubble volume is [14]

$$dV = 2\pi DR dR \tag{2.56}$$

During the bubble expansion, the work is taken as [14]

$$dW = \left(p_g + p_v - p_\infty\right)dV \tag{2.57}$$

where "p_g" represents the gas pressure, "p_v" represents the vapor pressure in the bubble. The potential energy associated with pressure "U_p" and surface tension "U_σ" can be obtained as [14]

$$dU_p = -2\pi D(p_g + p_v - p_\infty)R dR \tag{2.58}$$

$$dU_\sigma = 2\pi\sigma(D + 2R)dR \tag{2.59}$$

$$dU = dU_p + dU_\sigma = -2\pi RD(p_g + p_v - p_\infty) + 2\pi\sigma(D + 2R) \tag{2.60}$$

When the bubble is at equilibrium, the equation of bubble dynamics becomes [14]

$$\left(R\ddot{R} + \dot{R}^2\right)\ln\frac{r_0}{R} - \frac{1}{2}\dot{R}^2 = \frac{p_{ext} - p_0}{\rho_0} \tag{2.61}$$

$$p_{ext} = \left(p_0 + \frac{\sigma}{R_0} + \frac{2\sigma}{D} - p_v\right)\left(\frac{R_0}{R}\right)^{2\kappa} - \frac{\sigma}{R} - \frac{2\sigma}{D} - \frac{2\mu}{R}\dot{R} + p_v \tag{2.62}$$

2.2.3 Kedrinskii Equation

The Kedrinskii equation for a cylindrical bubble can be derived from the Bernoulli equation, where the Bernoulli equation is given as [15]

$$-\frac{\partial\varphi}{\partial t} + \frac{1}{2}u^2 + h = 0 \tag{2.63}$$

Through Eq. (2.41), one can derive [15]

$$\frac{\partial\rho}{\partial t} + \rho\frac{\partial u}{\partial r} + u\frac{\partial\rho}{\partial r} + \frac{\rho u}{r} = 0 \tag{2.64}$$

$$\frac{1}{c^2}\frac{\partial h}{\partial t} + \frac{\partial u}{\partial r} + \frac{u}{c^2}\frac{\partial h}{\partial r} + \frac{u}{r} = 0 \tag{2.65}$$

Replacing "h" in Eq. (2.65) with Eqs. (2.64), (2.66) can be written as [15]

$$\frac{1}{c^2}\frac{\partial^2\varphi}{\partial t^2} - \frac{u}{c^2}\frac{\partial u}{\partial t} + \frac{\partial u}{\partial r} + \frac{u}{c^2}\frac{\partial^2\varphi}{\partial t\partial r} - \frac{u^2}{c^2}\frac{\partial u}{\partial r} + \frac{u}{r} = 0 \tag{2.66}$$

Placing "$u = -\partial\varphi/\partial r$", into Eqs. (2.66), (2.66) can be written as [15]

$$\frac{1}{c^2}\frac{\partial^2 \varphi}{\partial t^2} + \frac{2u}{c^2}\frac{\partial^2 \varphi}{\partial t \partial r} - \left(1 - \frac{u^2}{c^2}\right)\frac{\partial^2 \varphi}{\partial r^2} - \frac{1}{r}\frac{\partial \varphi}{\partial r} = 0 \qquad (2.67)$$

When the perturbation is weak enough, the "u/c" term in Eq. (2.67) can be neglected according to the acoustic approximation, so it can be expressed as [15]

$$\frac{1}{c_\infty^2}\frac{\partial^2 \varphi}{\partial t^2} - \frac{\partial^2 \varphi}{\partial r^2} - \frac{1}{r}\frac{\partial \varphi}{\partial r} = 0 \qquad (2.68)$$

Substituting "$\Phi = r^{1/2}\varphi$" into Eq. (2.66), we can obtain [15]

$$\frac{1}{c_\infty^2}\frac{\partial^2 \Phi}{\partial t^2} - \frac{\partial^2 \Phi}{\partial r^2} - \frac{1}{4r^2}\Phi = 0 \qquad (2.69)$$

When "$\Phi/4r^2$" is relatively small and can be neglected, Eq. (2.69) can be simplified as [15]

$$\frac{1}{c_\infty^2}\frac{\partial^2 \Phi}{\partial t^2} - \frac{\partial^2 \Phi}{\partial r^2} = 0 \qquad (2.70)$$

There is one solution to the wave Eq. (2.70) expressed as [15]

$$\Phi = \Phi(\xi) \qquad (2.71)$$

For an outward propagating wave, the conditions can be assumed to be [15]

$$\xi = t - r/c_\infty \qquad (2.72)$$

By introducing "$G = r^{1/2}(h + u^2/2)$" and Eqs. (2.63) and (2.72), one can obtain [15]

$$\frac{\partial \Phi}{\partial t} = \frac{\partial \Phi}{\partial \xi} = G \qquad (2.73)$$

According to Eqs. (2.72) and (2.73), we can obtain [15]

$$\frac{\partial \Phi}{\partial r} = \frac{\partial \Phi}{\partial \xi} \cdot \frac{\partial \xi}{\partial r} = -\frac{1}{c_\infty}G \qquad (2.74)$$

Substituting Eqs. (2.73) and (2.74) into Eq. (2.70), one can obtain [15]

$$\frac{\partial G}{\partial t} + c_\infty \frac{\partial G}{\partial r} = 0 \qquad (2.75)$$

According to the Kirkwood-Bethe equation [15], we can have [15]

$$r^{1/2}\frac{\partial h}{\partial t} + r^{1/2}u\frac{\partial u}{\partial t} + (c+u)\cdot\left(\frac{1}{2}r^{-1/2}(h+u^2/2) + r^{1/2}\left(\frac{\partial h}{\partial r} + u\frac{\partial u}{\partial r}\right)\right) = 0$$

$$(2.76)$$

By multiplying both the left and right sides of Eq. (2.76) by "$r^{1/2}$", Eq. (2.76) can be expressed as [15]

$$r\frac{\partial h}{\partial t} + ru\frac{\partial u}{\partial t} + \frac{1}{2}(c+u)h + \frac{1}{2}(c+u)\frac{u^2}{2}$$
$$+ rc\frac{\partial h}{\partial r} + rcu\frac{\partial u}{\partial r} + ru\frac{\partial h}{\partial r} + ru^2\frac{\partial u}{\partial r} = 0 \qquad (2.77)$$

Based on the definition of the derivative after the motion, we can obtain [15]

$$\frac{dh}{dt} = \frac{\partial h}{\partial t} + u\cdot\frac{\partial h}{\partial r} \qquad (2.78)$$

$$\frac{du}{dt} = \frac{\partial u}{\partial t} + u\cdot\frac{\partial u}{\partial r} \qquad (2.79)$$

Then, Eq. (2.77) can be simplified as [15]

$$r\frac{dh}{dt} + ru\frac{du}{dt} + \frac{1}{2}(c+u)h + \frac{1}{2}(c+u)\frac{u^2}{2}$$
$$+ rc\frac{\partial h}{\partial r} + rcu\frac{\partial u}{\partial r} = 0 \qquad (2.80)$$

According to Eq. (2.67), we can get [15]

$$\frac{\partial u}{\partial r} = -\frac{1}{c^2}\frac{\partial h}{\partial t} - \frac{u}{c^2}\frac{\partial h}{\partial r} - \frac{u}{r} = -\frac{1}{c^2}\frac{dh}{dt} - \frac{u}{r} \qquad (2.81)$$

$$\frac{\partial h}{\partial r} = -\frac{\partial u}{\partial t} - u\frac{\partial u}{\partial r} = -\frac{du}{dt} \qquad (2.82)$$

By substituting Eqs. (2.81) and (2.82) into Eq. (2.80), we can obtain [15]

$$r\left(1-\frac{u}{c}\right)\frac{du}{dt} + \frac{3}{4}\left(1-\frac{u}{3c}\right)u^2 = \frac{1}{2}\left(1+\frac{u}{c}\right)h + \frac{r}{c}\left(1-\frac{u}{c}\right)\frac{dh}{dt} \qquad (2.83)$$

Based on the kinematic boundary conditions and the enthalpic boundary conditions, one can derive [15]

$$h(R, t) = h_B(t) \qquad (2.84)$$

$$h_B = \frac{n}{n-1} \frac{p_\infty + B}{\rho_\infty} \left[\left(\frac{p_B + B}{p_\infty + B} \right)^{\frac{n-1}{n}} - 1 \right] \tag{2.85}$$

$$c_B^2 = c_\infty^2 + (n-1)h_B \tag{2.86}$$

$$R\left(1 - \frac{\dot{R}}{c_B}\right)\ddot{R} + \frac{3}{4}\left(1 - \frac{\dot{R}}{3c_B}\right)\dot{R}^2 = \frac{1}{2}\left(1 + \frac{\dot{R}}{c_B}\right)h_B$$
$$+ \frac{R}{c_B}\left(1 - \frac{\dot{R}}{c_B}\right)\frac{dh_B}{dt} \tag{2.87}$$

In summary, the bubble dynamics can be expressed as [15]

$$R\left(1 - \frac{\dot{R}}{c_B}\right)\ddot{R} + \frac{3}{4}\left(1 - \frac{\dot{R}}{3c_B}\right)\dot{R}^2 = \frac{1}{2}\left(1 + \frac{\dot{R}}{c_B}\right)h_B$$
$$+ \frac{R}{c_B}\left(1 - \frac{\dot{R}}{c_B}\right)\frac{dh_B}{dt} \tag{2.88}$$

In the case of incompressible fluids, Eq. (2.87) can be replaced as [16]

$$R\ddot{R} + \frac{3}{4}\dot{R}^2 = \frac{1}{2}h_B \tag{2.89}$$

2.3 Bubble in Droplet

Assume that the N–S equation for motion in the r direction in a Newtonian fluid is [17]

$$\frac{\partial u}{\partial t} + u\frac{\partial u}{\partial r} - \nu\left[\frac{1}{r^2}\frac{\partial}{\partial r}\left(r^2\frac{\partial u}{\partial r}\right) - \frac{2u}{r^2}\right] = -\frac{1}{\rho}\frac{\partial P}{\partial r} \tag{2.90}$$

By substituting Eq. (2.2) with Eq. (2.90), it can be obtained that [17]

$$\frac{R_b^2}{r^2}\frac{d^2R_b}{dt^2} + \frac{2R_b}{r^2}\left(\frac{dR_b}{dt}\right)^2 - \frac{2R_b^4}{r^5}\left(\frac{dR_b}{dt}\right)^2 = -\frac{1}{\rho}\frac{\partial P}{\partial r} \tag{2.91}$$

Integrating from the cavitation bubble surface "R_b" to the droplet surface "R_d" gives [17]

$$\int_{R_b}^{R_d} \frac{1}{r^2}dr = \frac{1}{R_b} - \frac{1}{R_d} \tag{2.92}$$

$$\int_{R_b}^{R_d} \frac{-2}{r^5} dr = \frac{1}{2R_d^4} - \frac{1}{2R_b^4} \tag{2.93}$$

Substituting Eqs. (2.92) and (2.93) into Eq. (2.91) results in [17]

$$\left(R_b - \frac{R_b^2}{R_d}\right)\frac{d^2R_b}{dt^2} + \left(\frac{3}{2} - \frac{2R_b}{R_d} + \frac{R_b^4}{2R_d^4}\right)\left(\frac{dR_b}{dt}\right)^2 = \frac{p_b - p_d}{\rho_L} \tag{2.94}$$

Equations (2.95) and (2.96) represent the radial combined force per unit area of the surface of the bubble outward and the pressure of the bubble on the gas side, respectively, and can be expressed as [17]

$$p_b = p_{\text{in}} - \frac{2\sigma}{R_b} - \frac{4\mu}{R_b}\frac{dR_b}{dt} \tag{2.95}$$

$$P_{\text{in}} = \left[P_0 + \frac{2\sigma}{R_{b0}}\left(1 + \frac{R_{b0}}{R_{d0}}\right)\right]\left(\frac{R_{b0}}{R_b}\right)^{3\kappa} \tag{2.96}$$

The radial synergistic force at the droplet surface and the gas pressure at the droplet-air junction can be expressed as [17]

$$p_d = p_{\text{out}} + \frac{2\sigma}{R_d} + \frac{4\mu_G}{R_d}\frac{dR_d}{dt} \tag{2.97}$$

$$P_{\text{out}} = P_0 + P_A \cos(\omega t + \delta) \tag{2.98}$$

Summarizing the above and substituting Eqs. (2.95) and (2.96) into Eq. (2.94) the expression of the R–P equation for a bubble in the droplet is given as [17]

$$\left(R_b - \frac{R_b^2}{R_d}\right)\frac{d^2R_b}{dt^2} + \left(\frac{3}{2} - \frac{2R_b}{R_d} + \frac{R_b^4}{2R_d^4}\right)\left(\frac{dR_b}{dt}\right)$$
$$= \frac{1}{\rho}\left[(p_{\text{in}} - p_{\text{out}}) - \left(\frac{2\sigma}{R_b} + \frac{2\sigma}{R_d}\right)\right.$$
$$\left. - \left(\frac{4\mu}{R_b}\frac{dR_b}{dt} + \frac{4\mu_G}{R_d}\frac{dR_d}{dt}\right)\right]^2 \tag{2.99}$$

Since the liquid is assumed to be incompressible, the volume of the liquid remains conserved during the oscillation process and relations can be written as [17]

$$\frac{4}{3}\pi R_d^3 - \frac{4}{3}\pi R_b^3 = \frac{4}{3}\pi R_{d0}^3 - \frac{4}{3}\pi R_{b0}^3 = const \tag{2.100}$$

where, "R_b" and "R_d" represents the instantaneous radius value at the interface between the bubble and the droplet, "R_{b0}"and "R_{d0}" represents the radius at the interface between the bubble and the droplet at the equilibrium state. Deriving and further simplifying Eqs. (2.99), (2.100) can be expressed as [17]

$$\frac{dR_d}{dt} = \frac{R_b^2}{R_d^2}\frac{dR_b}{dt} \tag{2.101}$$

Substituting Eq. (2.100) into Eq. (2.98), the force equilibrium relation equation for the cavitation bubble inside the droplet is expressed as [17]

$$\left(R_b - \frac{R_b^2}{R_d}\right)\frac{d^2R_b}{dt^2} + \left(\frac{3}{2} - \frac{2R_b}{R_d} + \frac{R_b^4}{2R_d^4}\right)\left(\frac{dR_b}{dt}\right)^2$$
$$= \frac{1}{\rho}\left[(p_{\text{in}} - p_{\text{out}}) - \frac{2\sigma}{R_b}\left(1 + \frac{R_b}{R_d}\right)\right.$$
$$\left. - \left(\frac{4\mu}{R_b} + \frac{4\mu_G R_b^2}{R_d^3}\right)\frac{dR_b}{dt}\right] \tag{2.102}$$

Substituting Eqs. (2.96) and (2.98) into Eq. (2.102) becomes [17]

$$\left(R_b - \frac{R_b^2}{R_d}\right)\frac{d^2R_b}{dt^2} + \left(\frac{4\mu}{\rho R_b} + \frac{4\mu_G R_b^2}{\rho_L R_d^3}\right)\frac{dR_b}{dt}$$
$$+ \left(\frac{3}{2} - \frac{2R_b}{R_d} + \frac{R_b^4}{2R_d^4}\right)\left(\frac{dR_b}{dt}\right)^2$$
$$= \frac{1}{\rho}\left\{\left[P_0 + \frac{2\sigma}{R_{b0}}\left(1 + \frac{R_{b0}}{R_{d0}}\right)\right]\left(\frac{R_{b0}}{R_b}\right)^{3\kappa}\right.$$
$$\left. - \frac{2\sigma}{R_b}\left(1 + \frac{R_b}{R_d}\right) - P_0 - P_A\cos(\omega t + \delta)\right\} \tag{2.103}$$

$$R_b^3 = R_d^3 - R_{d0}^3 + R_{b0}^3 \tag{2.104}$$

References

1. Brennen CE (2014) Cavitation and bubble dynamics. Cambridge University Press, Cambridge
2. Zheng X, Wang X, Zhang Y (2022) A single oscillating bubble in liquids with high Mach number. Ultrason Sonochem 85:105985
3. Cole RH (1948) Underwater explosions
4. Lezzi A, Prosperetti A (1987) Bubble dynamics in a compressible liquid. Part 2. Second-order theory. J Fluid Mech 185:289–321

5. Rayleigh L (1917) On the pressure developed in a liquid during the collapse of a spherical cavity. Philos Mag Ser 6(34):94–98
6. Plesset MS (1949) The dynamics of cavitation bubbles
7. Prosperetti A, Lezzi A (1986) Bubble dynamics in a compressible liquid. Part 1. First-order theory. J Fluid Mech 168:457–478
8. Keller JB, Miksis M (1980) Bubble oscillations of large amplitude. J Acoust Soc Am 68(2):628–633
9. Herring C (1941) Theory of the pulsations of the gas bubble produced by an underwater explosion. Columbia Univ., Division of National Defense Research
10. Tomita Y, Shima A (1977) On the behavior of a spherical bubble and the impulse pressure in a viscous compressible liquid. Bull JSME 20(149):1453–1460
11. Fujikawa S, Akamatsu T (1980) Effects of the non-equilibrium condensation of vapour on the pressure wave produced by the collapse of a bubble in a liquid. J Fluid Mech 97(3):481–512
12. Yasui K (2018) Acoustic cavitation and bubble dynamics. Springer, Cham
13. Landau LD, Lifshitz EM (2013) Course of theoretical physics. Elsevier, Amsterdam
14. Ilinskii YA, Zabolotskaya EA, Hay TA et al (2012) Models of cylindrical bubble pulsation. J Acoust Soc Am 132(3):1346–1357
15. Kedrinskiĭ VK (2005) Hydrodynamics of explosion: experiments and models. Springer, New York
16. Xi X, Bian H, Wang X, Zhang Y (2022) Review on dynamics of the cylindrical cavitation bubble, nuclear science and technology
17. Li Z, Wang X, Shen J et al (2024) Cavity dynamics and splashing mechanism in droplets. Springer, New York

Chapter 3
Bubble Oscillation Dynamics

In this chapter, the radial oscillation characteristics of the bubble are discussed. Firstly, the basic principles and solution processes of three analytical methods, namely the perturbation method, the multi-scale method and the Laplace transform method, are introduced respectively, which are employed to linearize and approximate solve the equations of bubble dynamics. On this basis, the oscillation characteristics of the bubble in the cases of free oscillation and driven oscillation are analyzed, respectively.

3.1 Perturbation Method

In this section, Eq. (3.1) for a cylindrical bubble is addressed exemplarily through the application of the perturbation method [1]:

$$\left(1 - \frac{\dot{R}}{c_l}\right)R\ddot{R} + \frac{3}{4}\left(1 - \frac{\dot{R}}{3c_l}\right)\dot{R}^2 = \frac{1}{2}\left(1 + \frac{\dot{R}}{c_l}\right)\frac{P_{ext}(R,t) - P_s(t)}{\rho_l}$$
$$+ \frac{R}{\rho_l c_l}\left(1 - \frac{\dot{R}}{c_l}\right)\frac{d[P_{ext}(R,t) - P_s(t)]}{dt} \quad (3.1)$$

According to the perturbation method, for a bubble oscillating with a small amplitude, its instantaneous radius can be approximately expressed as [1]

$$R = R_0\left(1 + x_1\varepsilon + x_2\varepsilon^2 + \cdots\right) \quad (3.2)$$

Here, "x_1" and "x_2" represent the non-dimensional perturbations of the radius "R", while "ε" represents a non-dimensional parameter that signifies the scale of the non-dimensional amplitude, and it is subject to the condition $0 < \varepsilon \ll 1$.

© The Author(s), under exclusive license to Springer Nature Switzerland AG 2024
X. Wang et al., *Fundamentals of Single Cavitation Bubble Dynamics*,
SpringerBriefs in Energy, https://doi.org/10.1007/978-3-031-75041-0_3

After incorporating Eq. (3.2) into Eq. (3.1) and applying the Taylor formula to the expressions "$\left(1 + x_1\varepsilon + x_2\varepsilon^2\right)^n$", we neglect all terms with "ε^2" and higher powers. Consequently, Eq. (3.1) can be expressed with precision up to the first order in "ε" in the following manner [1]

$$\ddot{x}_1 + 2\beta_{tot}\dot{x} + \omega_0^2 x_1 = 0 \tag{3.3}$$

where

$$\beta_{tot} = \beta_{vis} + \beta_{th} + \beta_{ac} \tag{3.4}$$

$$\beta_{vis} = \frac{3\mu_l}{4\rho_l R_0^2 M} \tag{3.5}$$

$$\beta_{th} = \frac{3\mu_{th}}{4\rho_l R_0^2 M} \tag{3.6}$$

$$\beta_{ac} = \frac{R_0}{c_l}\omega_0^2 \tag{3.7}$$

$$\omega_0^2 = \frac{1}{2M\,\rho_l R_0^2}\left[2\kappa P_0 + \frac{(2\kappa - 1)\sigma}{R_0}\right] \tag{3.8}$$

Here, "β_{tot}" represents the total damping constant. "β_{vis}" represents the viscous damping constant. "β_{th}" represents the thermal damping constant. "β_{ac}" represents the acoustic damping constant. "ω_0" denotes the natural frequency of the gap bubble. "β_{vis}", "β_{th}", and "β_{ac}" describe the energy decay resulting from liquid viscosity, heat exchange between the bubble and the surrounding fluid, and the acoustic propagation of the bubble's vibrations, respectively. For the bubble size under discussion, its duration is approximately in the microsecond range, and the bubble's oscillations are modeled as an adiabatic phenomenon. Consequently, the thermal damping coefficient β_{th} is disregarded due to "$\mu_{th} = 0$", and "ω_0" represents a particular frequency that is defined based on the inherent characteristics of the bubble–liquid interaction.

The solution to Eq. (3.3) can be represented as [1]

$$x_1 = e^{-\beta_{tot}t}A_1\cos(\omega_1 t + \delta_1) \tag{3.9}$$

where

$$\omega_1 = \sqrt{\omega_0^2 - \beta_{tot}^2} \tag{3.10}$$

$$A_1 = C_0\frac{\omega_0}{\omega_1} \tag{3.11}$$

$$\delta_1 = -\arctan\left(\frac{\beta_{tot}}{\sqrt{\omega_0^2 - \beta_{tot}^2}}\right) \tag{3.12}$$

"A_1" represents the magnitude of the first-order solution. "δ_1" denotes the initial phase for the first-order solution. "C_0" signifies the starting value of the "x_1", which must be predetermined.

Introducing Eq. (3.2) into Eq. (3.1) and only considering the term containing "ε^2", Eq. (3.1) correct to the second order of "ε" can be written as [1]

$$\ddot{x}_2 + 2\beta_{tot}\dot{x}_2 + \omega_0^2 x_2 = \frac{e^{-2\beta_{tot}t}\varphi_0}{R_0^2 M} \tag{3.13}$$

where

$$M = 1 + \frac{R_0}{c_l}\frac{3(\mu_{th} + \mu_l)}{\rho_l R_0^2} \tag{3.14}$$

$$\beta_{tot} = \beta_{vis} + \beta_{th} + \beta_{ac} \tag{3.15}$$

$$\beta_{vis} = \frac{3\mu_l}{4\rho_l R_0^2 M} \tag{3.16}$$

$$\beta_{th} = \frac{3\mu_{th}}{4\rho_l R_0^2 M} \tag{3.17}$$

$$\beta_{ac} = \frac{R_0}{c_l}\omega_0^2 \tag{3.18}$$

$$\omega_0^2 = \frac{1}{2M\,\rho_l R_0^2}\left[2\kappa P_0 + \frac{(2\kappa - 1)\sigma}{R_0}\right], \tag{3.19}$$

$$\varphi_0 = D_1 \cos(2\omega_1 t) + D_2 \sin(2\omega t) + D_3, \tag{3.20}$$

$$\begin{aligned} D_1 &= \left[\varphi_1 + \varphi_2\left(\beta_{tot}^2 - \omega_1^2\right) + \varphi_3\beta_{tot}\right.\\ &\quad + \left.\varphi_4\left(3\beta_{tot}\omega_1^2 - \beta_{tot}^3\right) + \varphi_5\left(\beta_{tot}^2 - \omega_1^2\right)\right]\cos(2\delta_1)\\ &\quad + \left[2\varphi_2\beta_{tot}\omega_1 + \varphi_3\omega_1 + \varphi_4\left(\omega_1^3 - 3\omega_1\beta_{tot}^2\right) + 2\varphi_5\beta_{tot}\omega_1\right]\sin(2\delta_1) \end{aligned} \tag{3.21}$$

$$\begin{aligned} D_2 &= \left[-\varphi_1 - \varphi_2\left(\beta_{tot}^2 - \omega_1^2\right) - \varphi_3\beta_{tot} - \varphi_4\left(3\beta_{tot}\omega_1^2 - \beta_{tot}^3\right)\right.\\ &\quad - \left.\varphi_5\left(\beta_{tot}^2 - \omega_1^2\right)\right]\sin(2\delta_1)\\ &\quad + \left[2\varphi_2\beta_{tot}\omega_1 + \varphi_3\omega_1 + \varphi_4\left(\omega_1^3 - 3\omega_1\beta_{tot}^2\right) + 2\varphi_5\beta_{tot}\omega_1\right]\cos(2\delta_1) \end{aligned} \tag{3.22}$$

$$D_3 = \varphi_1 + \varphi_2\left(\beta_{tot}^2 + \omega_1^2\right) + \varphi_3\beta_{tot}$$
$$- \varphi_4\left(\beta_{tot}\omega_1^2 + \beta_{tot}^3\right) + \varphi_5\left(\beta_{tot}^2 - \omega_1^2\right) \tag{3.23}$$

$$\varphi_1 = \frac{A_1^2}{2}\left[\frac{\left(2\kappa^2 + \kappa\right)P_0 R_0 + \left(2\kappa^2 + \kappa - 1\right)\sigma}{2\rho_l R_0}\right] \tag{3.24}$$

$$\varphi_2 = \frac{A_1^2}{2}\left[\frac{2\kappa P_0 R_0^2}{\rho_l c_l^2} + \frac{(2\kappa - 1)R_0\sigma}{\rho_l c_l^2} + \frac{3(\mu_{th} + \mu_l)R_0}{2\rho_l c_l} - \frac{3R_0^2}{4}\right] \tag{3.25}$$

$$\varphi_3 = -\frac{A_1^2}{2}\left[\frac{3(\mu_{th} + \mu_l)c_l + \left(8\kappa^2 - 2\kappa\right)P_0 R_0 + \left(8\kappa^2 - 2\kappa - 1\right)\sigma}{2\rho_l c_l}\right] \tag{3.26}$$

$$\varphi_4 = \frac{A_1^2}{2}\left[\frac{3(\mu_{th} + \mu_l)R_0^2 + \rho_l c_l R_0^3}{\rho_l c_l^2}\right] \tag{3.27}$$

$$\varphi_5 = -\frac{R_0^2 A_1^2}{2} \tag{3.28}$$

Here, "$D_1 - D_3$" and "$\varphi_0 - \varphi_5$" are employed in the process of auxiliary computations. According to Eqs. (3.14)–(3.19), β_{tot} and ω_0 in Eq. (3.13) are identical to those present in Eq. (3.3).

So, the solution of Eq. (3.13) can be denoted as [1]

$$x_2 = e^{-2\beta_{tot}t}[A_2\cos(2\omega_1 t + \delta_2) + B_2] \tag{3.29}$$

where

$$A_2 = \left|\pm\left[\frac{1}{R_0^4 M^2}\frac{D_1^2 + D_2^2}{\left(\omega_0^2 - 4\omega_1^2\right)^2 + 16\omega_1^2\beta_{tot}^2}\right]^{\frac{1}{2}}\right| \tag{3.30}$$

$$\delta_2 = \tan^{-1}\left[\frac{4D_1\omega_1\beta_{tot} - D_2\left(\omega_0^2 - 4\omega_1^2\right)}{4D_2\omega_1\beta_{tot} + D_1\left(\omega_0^2 - 4\omega_1^2\right)}\right] \tag{3.31}$$

$$B_2 = \frac{D_3}{\omega_0^2 R_0^2 M} \tag{3.32}$$

Here, "A_2" is the magnitude of the second-order solution. "δ_2" denotes the initial phase for the second-order solution. "B_2" is the constant term for the second-order solution.

By substituting Eqs. (3.9) and (3.29) into Eq. (3.2), we obtain the approximate analytical solution for the gap bubble oscillation equation, which boasts second-order accuracy [1]

$$R = R_0\{1 + \varepsilon e^{-\beta_{tot}t}A_1\cos(\omega_1 t + \delta_1) + \varepsilon^2 e^{-2\beta_{tot}t}[A_2\cos(2\omega_1 t + \delta_2) + B_2]\}$$
$$\tag{3.33}$$

3.2 Multi-scale Method

In this section, Eq. (3.34) for a cylindrical bubble is addressed exemplarily through the application of the multiscale method [2]:

$$R\ddot{R} + \frac{3}{4}\dot{R}^2 = \frac{1}{2}\frac{P_R - P_s}{\rho_l}, \tag{3.34}$$

where

$$P_R = P_{in} - \frac{\sigma}{R} - \frac{3\mu}{R}\dot{R}, \tag{3.35}$$

$$P_s = P_0 + P_A\cos(\Omega t), \tag{3.36}$$

$$P_{in} = P_{in,eq}\left(\frac{R_0}{R}\right)^{2\kappa}, \tag{3.37}$$

$$P_{in,eq} = P_0 + \frac{\sigma}{R_0}. \tag{3.38}$$

Here, "P_{in}" represents the pressure of the gas at the bubble interface, "$P_{in,eq}$" denotes the gas equilibrium pressure at the same interface, "σ" is the surface tension coefficient, "μ" is the liquid viscosity, "P_0" is the ambient pressure of the liquid, "R_0" is the bubble equilibrium radius, and "κ" is the polytropic exponent.

In accordance with the term "$\left(P_{in,eq}/\rho_l\right)^{1/2}/R_0$", which shares the same dimensions as "Ω", the following dimensionless parameters are defined as [2]

$$\tau = t\frac{\left(\frac{P_{in,eq}}{\rho_l}\right)^{1/2}}{R_0}, \tag{3.39}$$

$$\beta = \frac{\mu}{2R_0\left(\rho_l P_{in,eq}\right)^{1/2}}, \tag{3.40}$$

$$\omega = R_0\left(\frac{\rho_l}{P_{in,eq}}\right)^{1/2}\Omega, \tag{3.41}$$

$$\varphi = \frac{\sigma}{R_0 P_{in,eq}} \tag{3.42}$$

$$\eta = \frac{P_A}{P_0} \tag{3.43}$$

$$\xi = \frac{P_A}{P_{in,eq}} = (1 - \varphi)\eta \tag{3.44}$$

where, "ω" represents the dimensionless frequency of the external acoustic excitation, while "η" denotes the dimensionless amplitude.

In scenarios where the acoustic excitation is intense, the bubble struggles to retain its initial cylindrical form during oscillations. However, in the current study, the intensity of the external acoustic excitation is assumed to be low enough to ensure that the bubble retains its cylindrical shape. Consequently, the radius of the bubble at any given moment within a confined area can be approximated by considering the following expression [2]:

$$R = R_0(1 + x) = R_0(1 + \varepsilon u) \tag{3.45}$$

Here, both variables "x" and "u" represent the dimensionless perturbations relative to the radius "R". The magnitude of "u" is of the order of 1, while "ε" is a dimensionless quantity that is much smaller than 1, specifically with ($0 < \varepsilon \ll 1$).

By incorporating Eqs. (3.39)–(3.45) into Eq. (3.34) and then applying the Taylor series expansion to the multi-power terms in the form of "$(1 + \varepsilon u)^n$", we derive the oscillation equation as [2]

$$\ddot{u} + \omega_0^2 u = -\frac{\xi}{2\varepsilon}\cos(\omega\tau) + \left[-\frac{3}{4}\dot{u}^2 + \frac{\xi}{2\varepsilon}u\cos(\omega\tau) - 3\frac{1}{\varepsilon}\beta\dot{u} + a_1 u^2\right]\varepsilon$$
$$+ \left[\frac{3}{4}u\dot{u}^2 - \frac{\xi}{2\varepsilon}u^2\cos(\omega\tau) + 6\frac{1}{\varepsilon}\beta u\dot{u} - a_2 u^3\right]\varepsilon^2 \tag{3.46}$$

where

$$\omega_0^2 = \kappa - \frac{1}{2}\varphi \tag{3.47}$$

$$a_1 = \frac{\left(2\kappa^2 + 3\kappa\right)}{2} - \varphi \tag{3.48}$$

$$a_2 = \frac{1}{6}\left(4\kappa^3 + 12\kappa^2 + 11\kappa\right) - \frac{3}{2}\varphi \tag{3.49}$$

$$\dot{u} = \frac{du}{d\tau} \tag{3.50}$$

$$\ddot{u} = \frac{d^2 u}{d\tau^2} \tag{3.51}$$

Here, "ξ" and "β" are considered to be small parameters, with their order being "ε^1" to the first power or higher. Additionally, "ω_0" represents the natural frequency of the bubble. The equations do not retain terms that are of the order "ε^3" or higher.

Based on the ratio of "ω" to "ω_0", the resonance phenomenon can be classified into three types: (i) primary resonance, where "$\omega \approx \omega_0$", (ii) superharmonic resonance, where "$\omega \approx 1/n \cdot \omega_0, n \in N$", (iii) subharmonic resonance, where "$\omega \approx n\omega_0, \ n \in N$". Each type of resonance leads to a distinct analytical solution for the oscillation equation. In the current study, the focus is exclusively on the primary resonance.

In a confined space, the oscillations of a bubble are constrained by factors of damping and nonlinearity. To effectively apply the multi-scale technique to the oscillation equation, it is essential to harmonize the terms associated with excitation, damping, and nonlinearity so that they are all incorporated into a single, consistent equation. Consequently, a detuning parameter, denoted by "δ", is introduced to signify the disparity between the excitation frequency ω and the natural frequency ω_0. This parameter is defined as [2]

$$\beta = \varepsilon^2 B \tag{3.52}$$

$$\xi = \varepsilon^3 P \tag{3.53}$$

$$\omega = \omega_0 + \varepsilon^2 \delta \tag{3.54}$$

where, the variables "B", "P", and "δ" are all of the same order of magnitude, typically 1. Consequently, the oscillation equation for the primary resonance of the bubble within a restricted space is derived as shown below [2]:

$$\ddot{u} + \omega_0^2 u = \left(-\frac{3}{4}\dot{u}^2 + a_1 u^2 \right)\varepsilon$$
$$+ \left[\frac{3}{4}u\dot{u}^2 - \frac{P}{2}\cos(\omega_0 \tau + \varepsilon^2 \delta \tau) - 3B\dot{u} - a_2 u^3 \right]\varepsilon^2 \tag{3.55}$$

The multi-scale method [3] uses a small parameter "ε" to create different time scales (like "t", "εt" and "$\varepsilon^2 t$") for a nonlinear equation. By separating the equation into parts with different scales, the method isolates and solves equations at each order of ε. Such treatment helps to simplify the problem and find approximate solutions on slower time scales.

In this section, under the assumption that the strength of the external acoustic excitation is significantly low, the multi-scale technique is applied to find the second-order approximation of the solution for Eq. (3.55). Using various time scales as independent variables, the term "u" within Eq. (3.55) can be expanded as [2]

$$u(\tau; \varepsilon) = u_0(T_0, T_1, T_2) + \varepsilon u_1(T_0, T_1, T_2) + \varepsilon^2 u_2(T_0, T_1, T_2) + \cdots \quad (3.56)$$

where

$$T_n = \varepsilon^n \tau; \quad (n = 0, 1, 2, \ldots) \quad (3.57)$$

The derivatives of first and second order with respect to the dimensionless time "τ" can be represented across multiple time scales in the following manner [2]

$$\frac{d}{d\tau} = D_0 + \varepsilon D_1 + \varepsilon^2 D_2 + \cdots \quad (3.58)$$

$$\frac{d^2}{d\tau^2} = D_0^2 + 2\varepsilon D_0 D_1 + \varepsilon^2 \left(2D_0 D_2 + D_1^2\right) + \cdots \quad (3.59)$$

where

$$D_n = \frac{\partial}{\partial T_n}, (n = 0, 1, 2, \ldots) \quad (3.60)$$

By inserting Eqs. (3.56)–(3.60) into Eq. (3.57) and matching the coefficients of "ε", expressions are derived that correspond to the orders of "ε^0", "ε^1", and "ε^2" as [2]

$$D_0^2 u_0 + \omega_0^2 u_0 = 0 \quad (3.61)$$

$$D_0^2 u_1 + \omega_0^2 u_1 = -\frac{3}{4}(D_0 u_0)^2 + a_1 u_0^2 - 2D_0 D_1 u_0 \quad (3.62)$$

$$D_0^2 u_2 + \omega_0^2 u_2 = \frac{3}{4}\left[u_0(D_0 u_0)^2 - 2D_0 u_0(D_1 u_0 + D_0 u_1)\right]$$
$$- \frac{P}{2} \cos(\omega_0 T_0 + \delta T_2) - 3B D_0 u_0 + 2a_1 u_0 u_1$$
$$- a_2 u_0^3 - 2D_0 D_2 u_0 - D_1^2 u_0 - 2D_0 D_1 u_1 \quad (3.63)$$

The solution to Eq. (3.61) is [2]

$$u_0 = A(T_1, T_2) \exp(i\omega_0 T_0) + c.c. \quad (3.64)$$

Here, "$c.c.$" denotes the complex conjugate of all preceding terms, and "A" signifies an unknown complex function that depends on the time scales "T_1" and "T_2".

Introducing Eq. (3.64) into Eq. (3.62), we obtain [2]

$$D_0^2 u_1 + \omega_0^2 u_1 = -2\frac{\partial A}{\partial T_1} i\omega_0 \exp(i\omega_0 T_0)$$

$$+ A^2 \left(a_1 + \frac{3}{4}\omega_0^2\right) \exp(2i\omega_0 T_0) + A\overline{A}\left(a_1 - \frac{3}{4}\omega_0^2\right) + c.c. \quad (3.65)$$

where "$\overline{A}$" is the complex conjugate of "A".

The infinite growth of the bubble amplitude over time is mitigated by removing the secular term from u_1. By making the term with "$\exp(i\omega_0 T_0)$" in Eq. (3.65) equal to zero, A is streamlined into a function of "T_2". As a result, the solution for Eq. (3.65) can be formulated as [2]

$$u_1 = m(T_2) \exp(2i\omega_0 T_0) + n(T_2) + c.c. \quad (3.66)$$

where

$$m = \frac{A^2}{-3\omega_0^2}\left(a_1 + \frac{3}{4}\omega_0^2\right) \quad (3.67)$$

$$n = \frac{A\overline{A}}{\omega_0^2}\left(a_1 - \frac{3}{4}\omega_0^2\right) \quad (3.68)$$

In order to determine the complex function, we insert Eqs. (3.64) and (3.66) into Eq. (3.63) to obtain [2]

$$D_0^2 u_2 + \omega_0^2 u_2 = \left[\frac{3}{4}A^2\overline{A}\omega_0^2 - 3\overline{A}\omega_0^2 m - \frac{P}{4}\exp(i\delta T_2) - 3BAi\omega_0\right.$$

$$+ 2a_1\left(\overline{A}m + An + A\overline{n}\right) - 3a_2 A^2\overline{A} - 2\frac{\partial A}{\partial T_2} i\omega_0 \bigg] \exp(i\omega_0 T_0)$$

$$+ \left(-\frac{3}{4}A^3\omega_0^2 + 3A\omega_0^2 m + 2a_1 Am - a_2 A^3\right)\exp(3i\omega_0 T_0) + c.c$$

$$(3.69)$$

Here, "$\overline{n}$" is the complex conjugate of "n".

Similarly, to eliminate the secular term in "u_2", set the term containing "$\exp(i\omega_0 T_0)$" in Eq. (3.69) to zero and introduce the parameter.

Likewise, to eradicate the secular component from "u_2", adjust Eq. (3.69) by making the expression with "$\exp(i\omega_0 T_0)$" equal to zero, and introduce the parameter [2]

$$A = \frac{1}{2}\alpha(T_2)\exp\{i[\delta T_2 - \gamma(T_2)]\} \quad (3.70)$$

Here, both "α" and "γ" are real-valued functions dependent on the time scale "T_2", Therefore, we obtain [2]

$$\frac{1}{8}\alpha^3 \left(\frac{10a_1^2}{3\omega_0^2} + \frac{3}{2}\omega_0^2 - \frac{5}{2}a_1 - 3a_2 \right) \exp\left[i(\delta T_2 - \gamma)\right]$$
$$- i\omega_0 \left[\frac{d\alpha}{dT_2} + \alpha i \left(\delta - \frac{d\gamma}{dT_2} \right) + \frac{3}{2}\alpha B \right] \exp\left[i(\delta T_2 - \gamma)\right]$$
$$- \frac{P}{4} \exp(i\delta T_2) = 0 \tag{3.71}$$

By applying Euler's formula and subsequently distinguishing between the real and imaginary components, Eq. (3.71) can be reformulated as [2]

$$\frac{d\alpha}{dT_2} = -\frac{P}{4\omega_0} \sin(\gamma) - \frac{3}{2}\alpha B \tag{3.72}$$

$$\frac{d\gamma}{dT_2} = \frac{\alpha^2 g_0}{8\omega_0} - \frac{P}{4\alpha\omega_0} \cos(\gamma) + \delta \tag{3.73}$$

where

$$g_0 = \left(\frac{10a_1^2}{3\omega_0^2} + \frac{3}{2}\omega_0^2 - \frac{5}{2}a_1 - 3a_2 \right) \tag{3.74}$$

Taking "$G = \varepsilon\alpha$" and introducing Eqs. (3.50)–(3.52) into Eqs. (3.72) and (3.73), we have [2]

$$\frac{dG}{d\tau} = -\frac{\xi}{4\omega_0} \sin(\gamma) - \frac{3}{2}\beta G \tag{3.75}$$

$$\frac{d\gamma}{d\tau} = \frac{G^2 g_0}{8\omega_0} - \frac{\xi}{4G\omega_0} \cos(\gamma) + \omega - \omega_0 \tag{3.76}$$

By merging Eqs. (3.64) and (3.66), the second-order approximation for the bubble oscillation equation can be expressed as [2]

$$\begin{aligned}
x = \varepsilon u &= \varepsilon u_0 + \varepsilon^2 u_1 \\
&= G \cos(\omega\tau - \gamma) \\
&\quad + G^2 \{g_1 + g_2 \cos[2(\omega\tau - \gamma)]\}
\end{aligned} \tag{3.77}$$

where

$$g_1 = \frac{1}{2\omega_0^2} \left(a_1 - \frac{3}{4}\omega_0^2 \right) \tag{3.78}$$

$$g_2 = -\frac{1}{6\omega_0^2} \left(a_1 + \frac{3}{4}\omega_0^2 \right) \tag{3.79}$$

Here, "G" represents the amplitude of the response, and γ denotes the phase of the approximate solution, both of which are crucial for the stability of the analytical result. By leveraging the connection between "G" and "α", we utilize "α" to signify the resonance frequency during the subsequent analytical examination.

As a bubble undergoes steady oscillations within a confined area, taking "$dG/d\tau = 0$" and "$d\gamma/d\tau = 0$", the frequency response equation can be formulated as [2]

$$\left[\left(\varepsilon^2\delta + \frac{g_0}{8\omega_0}G^2\right)^2 + \frac{9}{4}\beta^2\right]G^2 = \frac{\xi^2}{16\omega_0^2} \tag{3.80}$$

This expression indicates the interplay between the response amplitude "G", the detuning parameter "δ", and the amplitude of the external acoustic excitation "ξ". By solving Eq. (3.80), which includes a term with the sixth power of "G", and then substituting the derived solution into Eq. (3.77), the steady-state solution for the oscillation response is derived.

3.3 Laplace Transform Method

Here, we treat the cavitation bubble as a vapor bubble. For single-frequency acoustic field-driven bubble oscillation, the radial motion of the bubble can be expressed by the classical Rayleigh-Plesset equation [4] if the compressibility and viscosity of the liquid are sufficiently small and negligible. In addition, the size of the bubble is considered to be much smaller than the wavelength of the sound wave, and the portion of the sound pressure inhomogeneity around the bubble is negligible for both simple sinusoidal and standing waves. On this basis, the bubble oscillation can be described as [4]

$$R\ddot{R} + \frac{3}{2}\dot{R}^2 = \frac{p_R - p_s(t)}{\rho} \tag{3.81}$$

where

$$p_R = p_v - \frac{2\sigma}{R} \tag{3.82}$$

$$p_s(t) = p_\infty + p_a \sin(\omega_a t) \tag{3.83}$$

Here, "p_v" is the vapor pressure inside the bubble; "p_a" is the pressure amplitude of the sound field; and "ω_a" is the angular frequency of the driving sound field.

Since this section focuses on the small-amplitude oscillations of the cavitation bubble in the early stage of oscillation, the complex nonlinear phenomenon is not considered. And further neglecting the oscillation loss, Eq. (3.81) can be linearized as follows [5]

$$\ddot{x} + \omega_0^2 x = f \tag{3.84}$$

where

$$x = \frac{R - R_0}{R_0} \tag{3.85}$$

$$f = -\frac{p_a}{\rho R_0^2} \sin(\omega_a t) \tag{3.86}$$

Here, "$\ddot{x}$" is the second-order derivative of "x" with respect to time, "ω_0" is the intrinsic frequency of the bubble oscillation, and "x" is a dimensionless perturbation of the instantaneous radius of the bubble.

In order to close the above model, an expression applied to the intrinsic frequency of the vapor bubble is introduced as [5]

$$\omega_0^2 = \frac{1}{\rho R_0^2}\left(S - \frac{2\sigma}{R_0}\right) \tag{3.87}$$

where

$$S = \frac{2h_{lv}^2 \rho_v^2}{\rho c_{pl} T_0} \cdot \frac{\left(\frac{\omega_a R_0^2}{2D_l}\right)^{\frac{3}{2}}}{\left(\sqrt{\frac{\omega_a R_0^2}{2D_l}} + 1\right)^2 + \frac{\omega_a R_0^2}{2D_l}} \tag{3.88}$$

Here "h_{lv}" is the latent heat of evaporation of the liquid; "ρ_v" is the density of the vapor in the cavitation bubble; "c_{pl}" is the specific heat of the liquid at a constant pressure. "T_0" is the ambient temperature; "D_l" is the thermal diffusivity of the liquid.

In this study, the Laplace transform is chosen to tackle the linearized differential equations describing bubble wall motion, as presented above. The standard steps for applying the Laplace transform to differential equations include: first, transforming the differential equation into an algebraic one using the Laplace transform; second, solving the resulting algebraic equation and factoring the outcome; third, applying the inverse Laplace transform to the factored result; and finally, performing the convolution operation on the non-homogeneous part of the differential equation to find the solution. Regardless of whether the system is at resonance or not, adhering to these steps allows for the derivation of the corresponding analytical solution, with the only distinction lying in the integrand forms during the convolution process. Consequently, the Laplace transform method is a suitable choice for finding analytical solutions in both resonance and non-resonance conditions for vapor bubbles driven by an acoustic field. The generalized approach to solving these equations will be outlined subsequently.

Initially, by applying the Laplace transform, we obtain [6]

$$\mathcal{L}\{\ddot{x} + \omega_0^2 x\} = \mathcal{L}\{f\} \tag{3.89}$$

Here, "$\mathcal{L}\{\}$" denotes the Laplace transform operation. Utilizing the principle of linearity of the Laplace transform, as referenced in [3], Eq. (3.89) can be expressed as [6]

$$\mathcal{L}\{\ddot{x}\} + \omega_0^2 \mathcal{L}\{x\} = \mathcal{L}\{f\} \tag{3.90}$$

Subsequently, by employing the differential rule of the Laplace transform [3], Eq. (3.90) can be rearranged in the following manner [6]

$$s^2 \mathcal{L}\{x\} - x_0 s - \dot{x}_0 + \omega_0^2 \mathcal{L}\{x\} = \mathcal{L}\{f\} \tag{3.91}$$

where, "x_0" and "$\dot{x}_0$" represent the initial values for the variable "x" and "$\dot{x}$" respectively. The solution to Eq. (3.91) is [6]

$$\mathcal{L}\{x\} = \frac{x_0 s + \dot{x}_0}{s^2 + \omega_0^2} + \frac{\mathcal{L}\{f\}}{s^2 + \omega_0^2} \tag{3.92}$$

Following this, by factoring Eq. (3.92), it is possible to derive the following expression [6].

$$\mathcal{L}\{x\} = \frac{x_0}{2}\left(\frac{1}{s - i\omega_0} + \frac{1}{s + i\omega_0}\right) + \frac{\dot{x}_0}{2i\omega_0}\left(\frac{1}{s - i\omega_0} - \frac{1}{s + i\omega_0}\right)$$
$$+ \frac{\mathcal{L}\{f\}}{2i\omega_0}\left(\frac{1}{s - i\omega_0} - \frac{1}{s + i\omega_0}\right) \tag{3.93}$$

In accordance with the exponential and linearity law of the Laplace transformation [4], Eq. (3.93) can be rewritten as [6]

$$\mathcal{L}\{x\} = \mathcal{L}\left\{\frac{x_0}{2}\left(e^{i\omega_0 t} + e^{-i\omega_0 t}\right)\right\} + \mathcal{L}\left\{\frac{\dot{x}_0}{2i\omega_0}\left(e^{i\omega_0 t} - e^{-i\omega_0 t}\right)\right\}$$
$$+ \mathcal{L}\{f\} \cdot \mathcal{L}\left\{\frac{1}{2i\omega_0}\left(e^{i\omega_0 t} - e^{-i\omega_0 t}\right)\right\} \tag{3.94}$$

Additionally, utilizing the convolution and linearity law of the Laplace transform [7], Eq. (3.94) may be represented in the following manner as [6]

$$\mathcal{L}\{x\} = \mathcal{L}\left\{x_0 \cdot \frac{e^{i\omega_0 t} + e^{-i\omega_0 t}}{2} + \dot{x}_0 \cdot \frac{e^{i\omega_0 t} - e^{-i\omega_0 t}}{2i\omega_0} + f * \frac{e^{i\omega_0 t} - e^{-i\omega_0 t}}{2i\omega_0}\right\} \tag{3.95}$$

Here, the symbol "$*$" represents the convolution operation. Owing to the uniqueness of the Laplace transform [4], the notation for the Laplace transform in Eq. (3.95) can be omitted directly. Consequently, we can derive [6]

$$x = x_0 \cdot \frac{e^{i\omega_0 t} + e^{-i\omega_0 t}}{2} + \dot{x}_0 \cdot \frac{e^{i\omega_0 t} - e^{-i\omega_0 t}}{2i\omega_0} + f * \frac{e^{i\omega_0 t} - e^{-i\omega_0 t}}{2i\omega_0} \tag{3.96}$$

Additionally, by applying Euler's formula [4], which states that "$e^{\pm i\omega_0 t} = \cos \omega_0 t \pm i \sin \omega_0 t$", we can deduce that Eq. (3.96) can be converted into [6]

$$x = x_0 \cdot \cos \omega_0 t + \frac{\dot{x}_0}{\omega_0} \cdot \sin \omega_0 t + f * \frac{\sin \omega_0 t}{\omega_0} \tag{3.97}$$

Finally, by performing the convolution operation on the final term of Eq. (3.97) and then merging Eq. (3.97), the outcome is achieved, expressed as [6]

$$x = x_0 \cdot \cos \omega_0 t + \frac{\dot{x}_0}{\omega_0} \cdot \sin \omega_0 t$$
$$- \frac{P_a}{\rho_l R_0^2} \frac{1}{\omega_0} \int_0^t \sin \omega_a \tau \cdot \sin \omega_0 (t - \tau) d\tau \tag{3.98}$$

Here, "τ" is an integral variable, and Eq. (3.98) provides the analytical form for the bubble's dimensionless radius.

In the resonance condition, the frequency "ω_0" is equal to "ω_a".

The value of "ω_a" is referred to as the resonance frequency of the oscillating vapor bubble, and it is designated as the new parameter "ω_r". "ω_0" changes with certain factors, allowing "ω_r" to be approximately determined using graphical methods.

In this case, utilizing the trigonometric product-to-sum identity, it is possible to derive [6]

$$\sin \omega_a \tau \cdot \sin \omega_0 (t - \tau) = \frac{1}{2}[\cos(2\omega_a \tau - \omega_a t) - \cos \omega_a t] \tag{3.99}$$

Subsequently, bring Eq. (3.99) into Eq. (3.98) and replace all instances of ω_0 with ω_a, thereby allowing the resonance solution for the vapor bubble's oscillation in the acoustic field to be stated as [6]

$$x_r = x_{r1} + x_{r2} + x_{r3} + x_{r4}, \quad \omega_0 = \omega_a \tag{3.100}$$

where

$$x_{r1} = x_0 \cdot \cos \omega_a t \tag{3.101}$$

$$x_{r2} = \frac{\dot{x}_0}{\omega_a} \cdot \sin \omega_a t \tag{3.102}$$

$$x_{r3} = -\frac{p_a}{\rho R_0^2} \frac{\sin \omega_a t}{2\omega_a^2} \tag{3.103}$$

$$x_{r4} = \frac{p_a}{\rho R_0^2} \frac{t \cos \omega_a t}{2\omega_a} \tag{3.104}$$

Here, "x_r" is the dimensionless perturbation of the instantaneous radius of the cavitation bubble in the resonance case; "$\ddot{x}$" is the second-order derivative of x with respect to time.

Equation (3.100) shows that the resonance solution "x_r" consists of four terms with different and well-defined physical significance. The first term "x_{r1}" is determined by the initial dimensionless perturbation of the radius of the bubble. The second term is determined by the initial rate of change of the dimensionless radius. The last two terms, "x_{r3}" and "x_{r4}", are both primarily related to the amplitude and frequency of the sound field. Nevertheless, the differences between x_{r3} and x_{r4} remain significant. In comparison, x_{r3} is bounded while x_{r4} is unbounded. For x_{r4}, its amplitude is a function of time and increases over time. Predictably, when the oscillation time is large enough, the amplitude of x_{r4} will be much larger than the other three bounded ones. Therefore, it can be assumed that x_{r4} is the root cause of the violent oscillations of the bubble.

In addition, the non-resonant solution for the oscillation of the bubble in the acoustic field is [6]

$$x_n = x_{n1} + x_{n2} + x_{n3} + x_{n4}, \quad \omega_0 \neq \omega_a \tag{3.105}$$

where

$$x_{n1} = x_0 \cdot \cos \omega_0 t \tag{3.106}$$

$$x_{n2} = \frac{\dot{x}_0}{\omega_0} \cdot \sin \omega_0 t \tag{3.107}$$

$$x_{n3} = -\frac{p_a}{\rho R_0^2} \frac{\sin \omega_a t + \sin \omega_0 t}{2\omega_0(\omega_a + \omega_0)} \tag{3.108}$$

$$x_{n4} = \frac{p_a}{\rho R_0^2} \frac{\sin \omega_a t - \sin \omega_0 t}{2\omega_0(\omega_a - \omega_0)} \tag{3.109}$$

Here, "x_n" is the dimensionless perturbation of the instantaneous radius of the bubble in the non-resonant case.

Figure 3.1 illustrates the oscillation properties of the cavitation bubble in the time and frequency domains in the case of resonance. In particular, Fig. 3.1a illustrates the oscillation properties of the dimensionless radius and its four terms in the time domain, and Fig. 3.1b illustrates the oscillation properties of the dimensionless radius and its four terms in the frequency domain. In Fig. 3.1a, the green dashed line, blue dotted line, orange dotted dashed line, and purple solid line correspond to the curves of different dimensionless perturbations of the resonant case, respectively, and the red thick solid line corresponds to the curve of the dimensionless radius x_r in the resonance case. In Fig. 3.1b, "A_r" denotes the amplitude of the cavitation bubble oscillation in the resonance case. As shown in Fig. 3.1a, the amplitude of oscillation of the dimensionless radius in the case of resonance increases gradually with time. Regarding its four terms, the curves always oscillate periodically with fixed amplitude, the same frequency and different phases. The curve corresponding to x_{r4}, on the other hand, has an amplitude that increases linearly with time and exhibits non-periodic oscillations. The oscillation process can be roughly divided into two phases based on the dominance of the four terms on x_r. Subsequently, the influence of each on x_r changes significantly from the initial stage. At this stage, x_{r4} becomes the dominant term and its dominant role increases with time because its amplitude increases linearly with time and is gradually larger than the amplitudes of others. As shown in Fig. 3.1b, there is only one characteristic angular frequency during the oscillation in the resonance case, which is equal to the resonance frequency ω_r.

Figure 3.2 demonstrates the oscillation properties of the cavitation bubble in the time and frequency domains for the non-resonant case. In particular, Fig. 3.2a illustrates the oscillation characteristics of the dimensionless radius and its four terms in the time domain, and Fig. 3.2b illustrates the oscillation characteristics of the dimensionless radius and its four terms in the frequency domain. In Fig. 3.2a, the green dashed line, blue dotted line, orange dotted dashed line, and violet solid line correspond to the different dimensionless perturbations in the non-resonant case, and the red thick solid line corresponds to the curve of x_n. In Fig. 3.2b, "A_n" denotes the amplitude of the cavitation bubble oscillation in the non-resonant case. As shown in

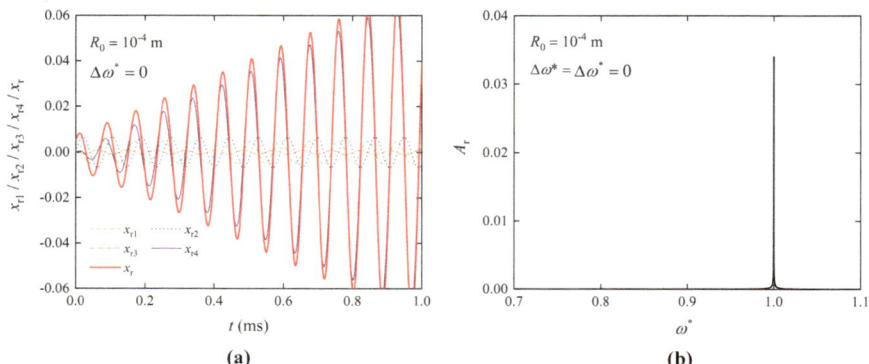

Fig. 3.1 Oscillation characteristics of the bubble in the time and frequency domains under resonance condition. **a** Time domain; **b** Frequency domain. Reprinted with the permission from Ref. [6] Copyright (2023) (Elsevier)

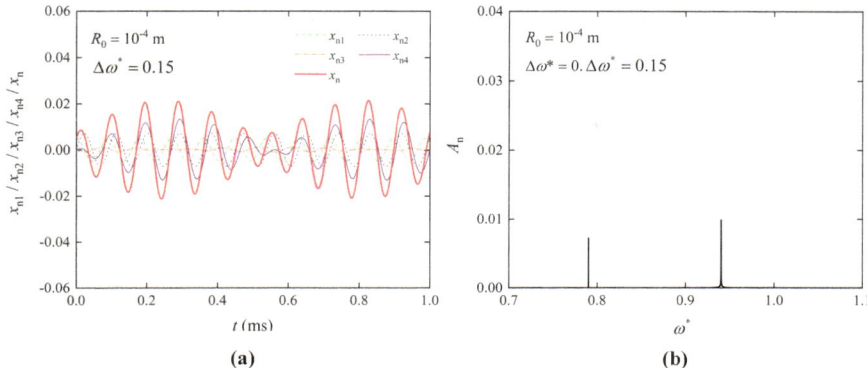

Fig. 3.2 Oscillation characteristics of the bubble in the time and frequency domains under non-resonance condition. **a** Time domain; **b** Frequency domain. Reprinted with the permission from Ref. [6] Copyright (2023) (Elsevier)

Fig. 3.2a, the amplitude of oscillation of the dimensionless radius increases periodically and then decreases in the nonresonant case. Among them, only the oscillation amplitude of the curve corresponding to x_{n4} exhibits obvious periodic changes over time, and always increases from zero to a fixed maximum value in a unit period, and then decreases back to zero. In other words, x_{n4} is the key reason for the periodic change in amplitude. Whereas the effects of others on the amplitude are always constant, these effects are mainly reflected at the beginning and end of each oscillation period of x_n. In general, the amplitude of the dimensionless radius cannot be increased indefinitely in the non-resonant case compared to the oscillation characteristics in the resonant case in the attached Fig. 3.2a. As shown in Fig. 3.2b, there are two distinct characteristic corner frequencies with different peak amplitudes.

3.4 Free Oscillation Characteristics of Bubbles

3.4.1 Weak Nonlinear Oscillation

This section demonstrates the characteristics of the weak nonlinear oscillations of bubbles by analyzing the damping constants and natural frequencies of the bubbles. Figure 3.3 illustrates the variation of the natural frequency of the bubble with the equilibrium radius of the bubble. The figure demonstrates that natural frequency experiences a notable decrease as the bubble equilibrium radius increases. This trend is consistent with Eq. (3.8), which defines natural frequency as a function of the bubble's equilibrium radius without considering variations in the liquid viscosity. Additionally, as the polytropic index increases, natural frequency gradually rises slightly, but the overall trend remains unchanged.

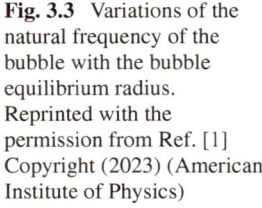

Fig. 3.3 Variations of the natural frequency of the bubble with the bubble equilibrium radius. Reprinted with the permission from Ref. [1] Copyright (2023) (American Institute of Physics)

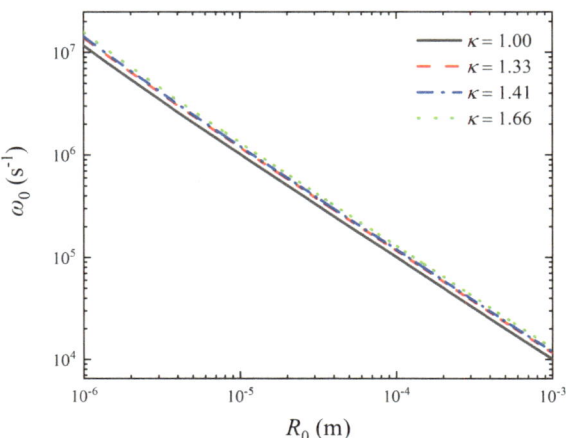

In Fig. 3.4, the relationship between the damping constants of a bubble and its equilibrium radius is illustrated. The blue dotted line corresponds to the acoustic damping constant, the red dashed line to the viscous damping constant, and the black solid line to the total damping constant. The light blue dashed lines denote the equilibrium radii at which the balance between acoustic and viscous damping shifts within the total damping. The right-hand scale of the figure displays the polytropic exponent for three different sets of damping constants, ranging polytropic exponent from 1.00 to 1.66 from left to right. The figure indicates that as R_0 increases, both acoustic damping constant and viscous damping constant decline, a trend that mirrors the decrease in total damping constant. Additionally, the viscous damping constant decreases at a greater rate than the acoustic damping constant, and the total damping's primary source switches from viscous damping constant to acoustic damping constant at its equilibrium radius is equal to 7.82×10^{-6} m.

Figure 3.5 illustrates how the relative difference in the natural frequency between spherical and cylindrical bubbles changes with the bubble equilibrium radius. The graph reveals that for bubble equilibrium radius values ranging from 10^{-6} to 10^{-3} m, the natural frequency of the spherical bubble is over 70% greater than that of the cylindrical bubble. As the bubble's equilibrium radius grows, the difference in natural frequency diminishes from 107.34 to 73.26%, and the curve representing this change becomes less steep and more gradual. When the bubble's equilibrium radius is beyond 10^{-4} m, the radius has a negligible impact on natural frequency.

Figure 3.6 depicts the trend of the relative differences in damping constants between spherical and cylindrical bubbles in relation to the bubble equilibrium radius. The blue dashed line, red dashed line and black solid line correspond to the relative differences in the acoustic damping constant, the viscous damping constant, and the total damping constant, respectively. According to Fig. 3.6, for the bubble's equilibrium radius values between 10^{-6} and 10^{-3} m, the damping constants of the spherical bubble are more than 50% greater than those of the cylindrical bubble. As the bubble's equilibrium radius increases, the changes in the acoustic damping constant,

Fig. 3.4 Variation of the damping constants of the bubble with the bubble equilibrium radius. Reprinted with the permission from Ref. [1] Copyright (2023) (American Institute of Physics)

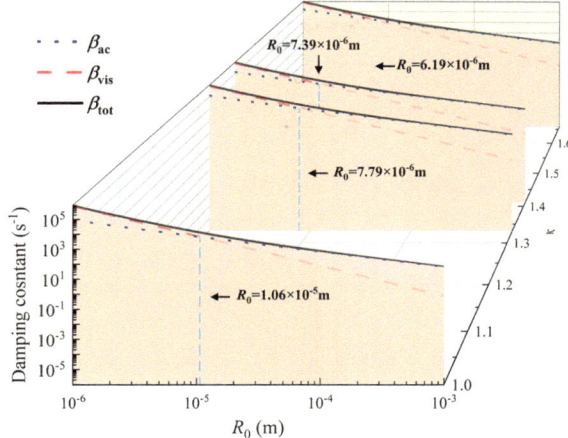

Fig. 3.5 Variation of the relative difference of the natural frequency between spherical and cylindrical bubbles with the bubble's equilibrium radius. Reprinted with the permission from Ref. [1] Copyright (2023) (American Institute of Physics)

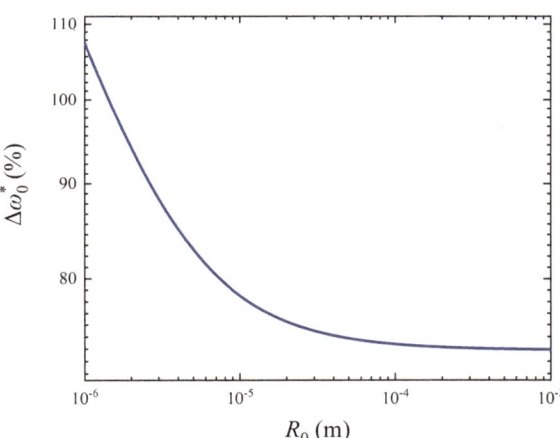

the viscous damping constant, and the total damping constant exhibit the following patterns: the acoustic damping constant and its rate of change both decrease gradually, reaching a stable state once the bubble's equilibrium radius exceeds 10^{-4} m. The viscous damping constant remains constant at 166.67%, while the total damping constant decreases over time, with its rate of change being slower than that of the acoustic damping constant.

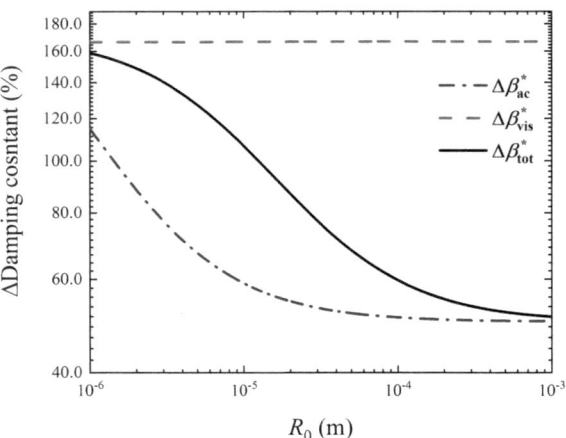

Fig. 3.6 Variation of the relative differences of damping constants for spherical and cylindrical bubbles with the bubble's equilibrium radius. Reprinted with the permission from Ref. [1] Copyright (2023) (American Institute of Physics)

3.4.2 Strong Nonlinear Oscillation

This section shows the difference in the behavior of the spherical and cylindrical bubbles in strong nonlinear oscillations with different parameters.

Figure 3.7 illustrates the evolution of the relative difference in the first peak of the bubble radius between spherical and cylindrical bubbles with respect to the dimensionless initial bubble radius. The black solid and dotted lines correspond to the first peak of the bubble radius bigger than zero and the first peak of the bubble radius smaller than zero, respectively. Two blue dashed lines are marked at the dimensionless initial bubble radius equal to 1.08 and 1.49. The blue point labeled A indicates the peak value of ΔR^*_{peak}. It is observed in Fig. 3.7 that as the dimensionless initial bubble radius grows, the first peak in the bubble radius initially increases slightly before decreasing, and its rate of change accelerates over time. At the dimensionless initial bubble radius is equal to 1.32, the first peak in the bubble radius reaches its maximum, which is 0.06%. Conversely, when the dimensionless initial bubble radius is equal to 3.0, the first peak in the bubble radius decreases to -14.67%.

Figure 3.8 depicts how the relative difference in the first peak of the bubble wall velocity changes with the dimensionless initial bubble radius. The blue point B marks the lowest point of the first peak of the bubble wall velocity. In Fig. 3.8, it is noted that the peak velocity of the spherical bubble consistently exceeds that of the cylindrical bubble, and the magnitude of the first peak of the bubble wall velocity is generally of the order of magnitude is 10^2. With an increase in dimensionless initial bubble radius, the first peak of the bubble wall velocity initially decreases slightly, reaching its minimum value of 87.99% at the dimensionless initial bubble radius is equal to 1.09, and then begins to rise.

Figure 3.9 presents the trend of the relative difference in the first peak of the bubble wall acceleration for spherical and cylindrical bubbles as a function of the dimensionless initial bubble radius. Two blue dotted lines, one horizontal and one

Fig. 3.7 Variation of the relative difference of the first peak in bubble radius between spherical and cylindrical bubbles with the dimensionless initial bubble radius. Reprinted with the permission from Ref. [1] Copyright (2023) (American Institute of Physics)

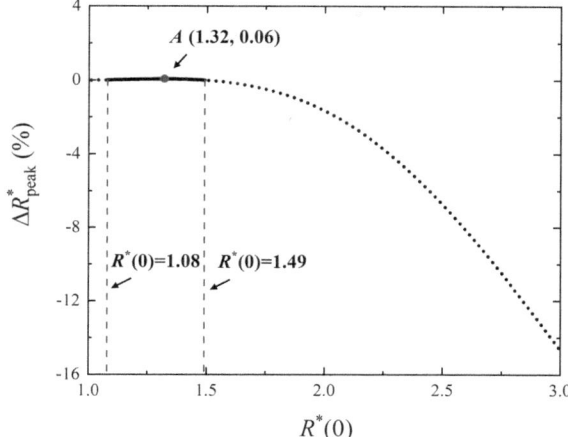

Fig. 3.8 Variation of the relative difference at the first peak of the bubble wall velocity between spherical and cylindrical bubbles with the dimensionless initial bubble radius. Reprinted with the permission from Ref. [1] Copyright (2023) (American Institute of Physics)

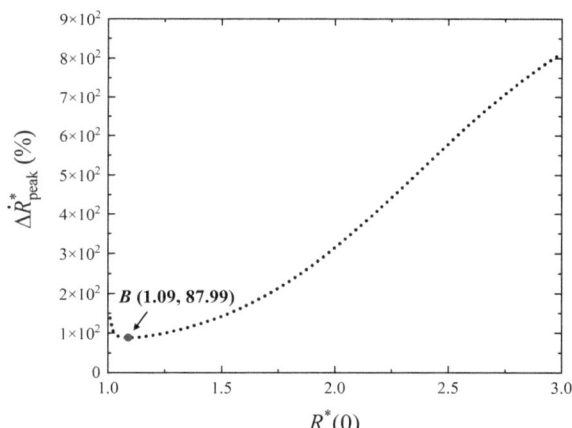

vertical, denote the first peak of the bubble wall acceleration is equal to $1.0 \times 10^{-4}\%$ and the dimensionless initial bubble radius is equal to 2.35, respectively. The blue point C indicates the threshold where the first peak of the bubble wall acceleration reaches $1.0 \times 10^{4}\%$. In Fig. 3.9, it is observed that the acceleration peak of the spherical bubble is consistently greater than that of the cylindrical bubble. As the dimensionless initial bubble radius increases, both the first peak of the bubble wall acceleration and its rate of change progressively rise. For the dimensionless initial bubble radius greater than 2.35, the magnitude of the first peak of the bubble wall acceleration is on the order of 10^{4}.

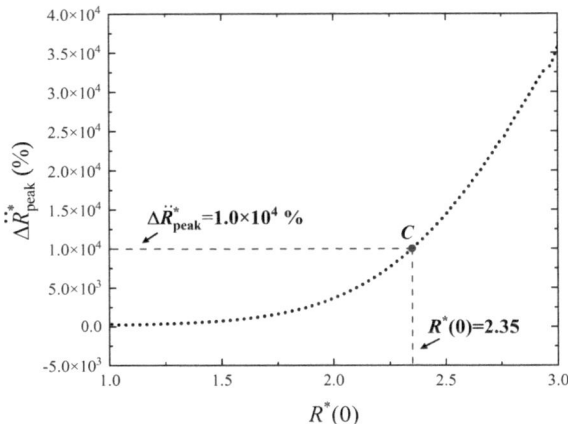

Fig. 3.9 Variation of the relative difference at the first peak of the bubble wall acceleration between spherical and cylindrical bubbles with the dimensionless initial bubble radius. Reprinted with the permission from Ref. [1] Copyright (2023) (American Institute of Physics)

3.5 Driven Oscillation Characteristics of Bubbles

This section discusses the oscillation characteristics of bubbles under single-frequency acoustic excitation. Figure 3.10 presents the frequency response diagram including amplitude versus frequency and phase versus frequency. Figure 3.10a displays the amplitude-frequency characteristic, with the red line depicting the response amplitude "α" as a function of the detuning parameter "δ". The trajectory of the dashed line indicates the unstable zone, and the black arrows illustrate the shifts in response amplitude with changes in the detuning parameter, accompanied by numbered nodes along these trends. The curve in 7(a) demonstrates a notable bending, with a pronounced peak, and exhibits the behavior of a soft spring. As the detuning parameter decreases, response amplitude rises from Point 1 to Point 2, then to Point 3, experiences a sudden drop to Point 4, and finally decreases to Point 5. Conversely, when the detuning parameter increases, response amplitude increases from Point 5 to Point 6, jumps to Point 2, and then descends to Point 7. This demonstrates that the path of response amplitude is significantly influenced by the direction of the detuning parameter increase or decrease, with different detuning parameter values corresponding to the "jumps" in response amplitude, indicating hysteresis. Moreover, response amplitude can have multiple corresponding values between Points 3 and 6, signifying an unstable region for the primary resonance solution of the bubble's oscillation equation. Figure 3.10b presents the phase-frequency characteristic, with the red line showing the response phase "γ" as a function of the detuning parameter. Similar to the amplitude-frequency curve, the response phase exhibits jump, hysteresis, and unstable regions. When the detuning parameter decreases the response phase, increases from Point 1 to Point 3, then jumps to Point 4. When the detuning parameter increases, the response phase decreases from Point 5 to Point 6, jumps down to Point 2, and continues to decrease to Point 7. The response phase can take on multiple values between Points 3 and 6. These trends fully elucidate the impact of the

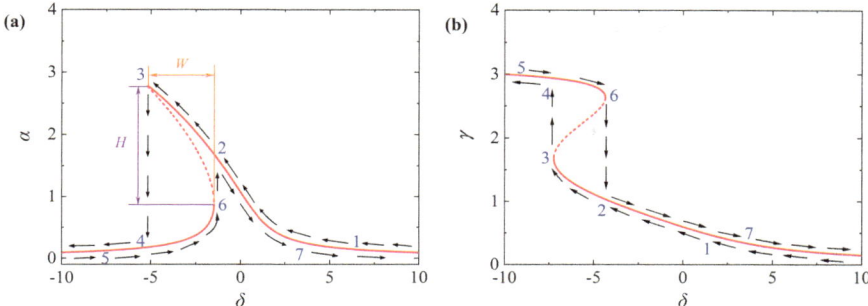

Fig. 3.10 Graphs of frequency responses. **a** The amplitude versus frequency plot. **b** The phase versus frequency plot. Reprinted with the permission from Ref. [2] Copyright (2024) (American Institute of Physics)

detuning parameter on the dimensionless analytical solution presented in Figs. 3.3 and 3.4 Given the similarity in the observed jumps, hysteresis, and multiple values across the graphs, the subsequent discussion will investigate the effects of external acoustic excitation and bubble equilibrium radius through the amplitude-frequency response curves.

Figure 3.11 shows the unstable region width and height with respect to the acoustic excitation amplitude. Figure 3.11a shows how the width of the unstable region, denoted as "D", changes in response to the amplitude of the external acoustic excitation. The black solid line, red dashed line, and blue dotted line correspond to cases with different initial bubble radii, respectively. With an increase in the acoustic excitation amplitude, the width of the unstable region expands at an increasingly faster rate. For higher values of the initial bubble radius, the width is larger and exhibits a greater responsiveness to variations in the acoustic excitation amplitude. When the initial bubble radius is equal to 1.0×10^{-5} m, and the acoustic excitation amplitude is equal to 0.005, the width of the unstable zone decreases to nearly zero. Figure 3.11b presents the height of the unstable region, labeled as "H", as it changes with the acoustic excitation amplitude. The same color-coded lines represent the same initial bubble radius values. The height rises approximately linearly with increasing acoustic excitation amplitude, and for larger values of the initial bubble radius, the height increases and its sensitivity to the acoustic excitation amplitude becomes slightly higher.

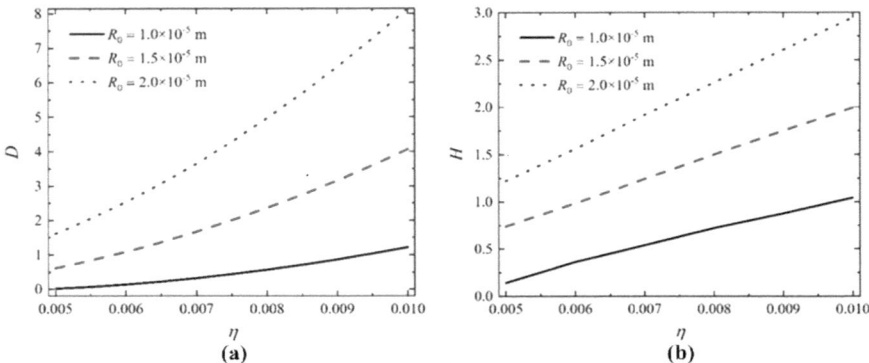

Fig. 3.11 Variation of the width and height of the unstable region with the acoustic excitation amplitude. Reprinted with the permission from Ref. [2] Copyright (2024) (American Institute of Physics)

References

1. Zhang X, Yang C, Wang C et al (2023) Dynamics of an oscillating cavitation bubble within a narrow gap. Phys Fluids 35(10):325
2. Wang X, Zhang X, Li S et al (2024) Primary resonance characteristics of a cylindrical bubble based on the multi-scale method. Phys Fluids 36(2):5481
3. Nayfeh AH (2008) Perturbation method. Wiley, Amsterdam
4. Khabeev NS (1981) Resonance properties of vapor bubbles. J Appl Math Mech 45(4):512–517
5. Prosperetti A (2015) The speed of sound in a gas–vapour bubbly liquid. Interface Focus 5(5):20150024
6. Wang X, Du X, Gao D et al (2023) Theoretical investigation on resonance characteristics of a vapor bubble based on Laplace transform method. Ultrason Sonochem 92:106275
7. Bronstein IN, Hromkovic J, Luderer B et al (2012) Taschenbuch der mathematik. Springer, New York

Chapter 4
Prediction Model of Bubble Collapse Jet

This chapter focuses on the Kelvin impulse theory model for predicting the jet intensity and direction induced by the bubble collapse. Firstly, the basic assumptions of the Kelvin impulse theory model, the derivation procedure and the Kelvin impulse formulae for spherical and cylindrical bubbles are presented. Secondly, the boundary treatment methods such as Weiss theorem, image method and conformal transformation are introduced. Finally, typical results of the Kelvin impulse with the bubbles located in different boundary conditions such as near a particle, near a wall with a bulge, inside a droplet and near a hydrofoil are described, respectively.

4.1 Kelvin Impulse

4.1.1 Basic Assumptions

The Kelvin impulse is an important physical quantity to describe the direction and strength of the jet induced by the bubble collapse. The theoretical model of the Kelvin impulse needs to meet the following assumptions:

(1) The liquid flow is considered as an incompressible potential flow. When the bubble is small, in the process of bubble collapse and jet formation, the maximum velocity of liquid around the bubble that can be attained is much smaller than the speed of sound in the water, and the pressure of the liquid is not enough to cause compression and density change of the liquid [1]. Therefore, the effects of liquid compressibility are neglected in the analysis. In addition, due to the weak vortex strength in the flow field induced by the bubble collapse, the flow of the liquid is treated as a potential flow to simplify the model.

© The Author(s), under exclusive license to Springer Nature Switzerland AG 2024
X. Wang et al., *Fundamentals of Single Cavitation Bubble Dynamics*,
SpringerBriefs in Energy, https://doi.org/10.1007/978-3-031-75041-0_4

(2) The effects of bubble oscillations on the liquid is considered as an isotropic point source with time-varying source strength and fixed position. Firstly, Best [2] discussed the plausibility of the spherical assumption of the bubble. During the entire period of the bubble growth and collapse, the bubble can remain quasi-spherical with a fixed position of the bubble centroid for the vast majority of the time. Significant non-spherical collapse occurs only 2–3% of the time at the end of the collapse period. Thus, the Kelvin impulse can be used to deduce subsequent deformation, movement and jets from the early behaviors of the bubble as it remains spherical. In addition, the influence of the "source" term tends to dominate in the case of bubble oscillations and boundary-induced perturbations to the surrounding liquid [3], while other higher-order terms associated with bubble movement and non-spherical deformation are neglected.

(3) The effects of buoyancy are neglected. When the bubble is small (millimeter scale), the buoyancy force is very small. According to the experimental results, when the bubble is located in a bulk liquid medium, the vertical movement distance of the bubble centroid is much smaller than the bubble's maximum radius during the bubble growth and collapse process. In addition, compared to the pressure gradient on the bubble surface caused by the boundary (generally characterized by the Bjerknes force), the effects of buoyancy are quite small, not exceeding 1% [4]. Therefore, the effect of buoyancy can be safely neglected.

4.1.2 Derivation Process

Firstly, Figs. 4.1 and 4.2 review scholars' contributions to the construction, development and application of the Kelvin impulse model, respectively. Among them, Fig. 4.1 presents the construction and refinement process of the Kelvin impulse. Figure 4.2 presents the application of Kelvin impulse for the bubble near different boundaries.

Then, Fig. 4.3 shows the derivation process of the Kelvin impulse model. Firstly, a liquid control body is selected, which is bounded by the bubble surface, the wall, and the liquid boundary at infinity. And the impulse relation equation of the control body is established. Based on the Reynolds transport theorem and the unsteady Bernoulli equation, the force exerted by the boundary on the control body is obtained. Next, according to the force balance between the inner and outer surfaces of the control body, the force exerted by the wall on the bubble surface is obtained, which is characterized by the Bjerknes force. Furthermore, the Lagally theorem applicable to deformable bodies in unsteady flows is introduced to simplify the Bjerknes force. Finally, the analytical form of the Kelvin impulse function can be established by integrating the Bjerknes force over time. In the function, the specific influence of the boundary on the bubble can be reflected by the term of additional velocity potential. In addition, to close the model, a suitable equation of bubble dynamics is introduced into the Kelvin impulse function to calculate the instantaneous radius of the bubble and its rate of change.

Scholars	Main contribution and key formulas
Benjamin and Ellis	First applied Kelvin impulse to collapsing jets. $$I = \int F \, dt$$
Blake et al. (1982)	Established the Kelvin impulse theory core system. $$I = \rho \int_{V'} u \, dV = \rho \int_{V'} \nabla \Phi \, dV = \rho \int_{S'} \Phi n \, dA$$
Blake et al. (1986-1988)	Extracted a function to represent different boundaries and applied to typical scenes. $$I_x(T_c) = \frac{2\pi\sqrt{6}R_m^5 \sqrt{\Delta p \rho}}{9h^2}\left[2\gamma^2\delta^2 B\left(\tfrac{11}{6},\tfrac{1}{2}\right)\cos\theta + \chi B\left(\tfrac{7}{6},\tfrac{3}{2}\right)\right]$$
Wang et al.(2023) and Shen et al.	Constructed a Kelvin theory model for cylindrical bubble. $$I = I_\Sigma + I_g$$ $$I_\Sigma = -\tfrac{16}{9}\sqrt{6}\pi h \rho^{1/2}\Delta p^{1/2}R_{\max}^3 g(s_0)B\left(\tfrac{3}{2},\tfrac{3}{2}\right)$$ $$I_g = \tfrac{2}{3}\sqrt{6}\rho^{3/2}\Delta p^{-1/2}g\pi h R_{\max}^3 B\left(\tfrac{5}{2},\tfrac{1}{2}\right)e_y$$
Wang et al. (2022)	Established an analytical expression for predicting the speed of motion of bubble collapse advection. $$v_{c.avg} = \frac{1}{\alpha}I$$ $$I = -\frac{8\sqrt{6}}{9}\pi\sqrt{(p_0 - p_v)\rho}R_{\max}^5 B\left(\tfrac{7}{6},\tfrac{3}{2}\right)\frac{R_p^3}{l^2(l^2 - R_p^2)^2}$$
Brujan et al.	Proposed the "zero final Kelvin pulse state". $$I = \frac{2}{3}\pi\left[R^3 U - \frac{3}{4}\left(\frac{R}{H}\right)^2 R^3 \dot{R}\right]$$
Supponen et al. and Andrews et al.	A parameter describing jets was extracted based on the Kelvin impulse. $$I_{surface} = 4.789R_0^3\sqrt{\Delta p \rho}\,\zeta n \times \begin{cases} -1 & \text{flat rigid surface} \\ +1 & \text{flat free surface} \end{cases}$$
Best and Blake	Obtained Kelvin impulse expression based on Lagally theorem. $$F_\Sigma = -4\pi\rho\sum_s P_q D_s^q \left(\nabla\phi'\right)_s$$ $$I = \pi\Gamma\int_0^{t'} R^4 \dot{R}^2 \, dt' = \pi\Gamma\int_0^R R'^4 \dot{R} \, dR'$$

Fig. 4.1 The construction and refinement of the Kelvin impulse [3–6, 10, 17–24]

Next, the derivation process of the Kelvin impulse will be introduced in detail. Blake and Cerone [5], Blake [6] systematically derived the formulae for the Kelvin impulse. Assuming that the initial momentum of the non-viscous and potential fluid with a volume of "V'" and a boundary surface of "S'" is zero. The impulse can be expressed as [5]

Scholars	Different boundary types
Wang et al. and Li et al. (2020, 2021)	Angular wall composed of two rigid walls
Wang et al. (2022)	Spherical particle
Li et al. (2023)	Angular wall composed by free surface and rigid wall
Wang et al. (2024)	Bulged right-angled walls at corner
Xu et al.	Vertical rigid wall
Wang et al. (2024)	Hemispherical bulge on flat wall
Ren et al.	Hemispherical oil droplet immersed in water

Fig. 4.2 The application of the Kelvin impulse [10, 15, 25–30]

$$I = \rho \int_{V'} u dV = \rho \int_{V'} \nabla \Phi dV = \rho \int_{S'} \Phi n dA \qquad (4.1)$$

where "I" is the Kelvin impulse. "ρ" represents the density of the liquid. "dV" represents the volume microelement. "u" represents the fluid local velocity. "Φ. " represents the fluid velocity potential. "n" represents the unit vector. "dA" represents an area microelement.

In the Kelvin impulse theory of a bubble near the boundary in a semi-infinite fluid, a control body of volume "V'" is selected as shown in Fig. 4.4, and its boundary "S'" consists of three parts, the surface of the bubble "S", the control surface of the flat boundary of the semi-infinite fluid "Σ_b", and the external boundary surface "Σ". The impulse of the control body can be expressed as the sum of two integrals with the following expression [5]

$$I = \rho \left\{ \int_S + \int_{\Sigma \cup \Sigma_b} \right\} \Phi n dA = I_S + I_{\Sigma \cup \Sigma_b} \qquad (4.2)$$

where "I_S" represents the impulse on the surface "S" of the bubble. "$I_{\Sigma \cup \Sigma_b}$" represents the impulse on the other boundary surfaces "$\Sigma \cup \Sigma_b$" of the control body.

For the finite volume "V", the change rate of fluid momentum is equal to the combined force acting on the control body, which can be expressed as [5]

$$F = \frac{dI}{dt} = -\left\{ \int_S + \int_{\Sigma \cup \Sigma_b} \right\} p n dA \qquad (4.3)$$

where "t" represents time; "p" represents the fluid pressure at a point.

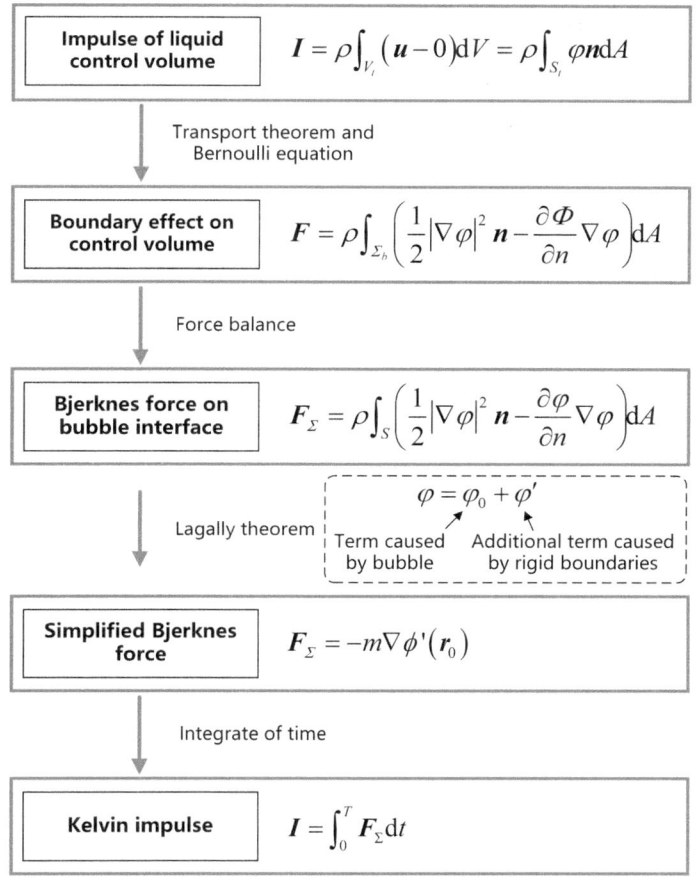

Fig. 4.3 Flow diagram of the Kelvin impulse derivation

Assuming a uniform distribution of vapor within the bubble, the pressure on the surface "*S*" of the bubble is uniform and approximately equal to the saturated vapor pressure within the bubble. And substituting Eq. (4.2) into Eq. (4.3), we obtain [5]

$$\frac{\mathrm{d}I_s}{\mathrm{d}t} = -\int_{\Sigma \cup \Sigma_b} pn\mathrm{d}A - \rho \frac{\mathrm{d}}{\mathrm{d}t} \int_{\Sigma \cup \Sigma_b} \Phi n\mathrm{d}A \tag{4.4}$$

The Reynolds transport theorem in a continuous medium can be expressed by Eq. (4.5) as [7, 8]

$$\frac{\mathrm{d}}{\mathrm{d}t} \int_{V'} f\mathrm{d}V = \int_{V'} \frac{\partial f}{\partial t}\mathrm{d}V + \int_{S'} (u \cdot n)f\mathrm{d}A \tag{4.5}$$

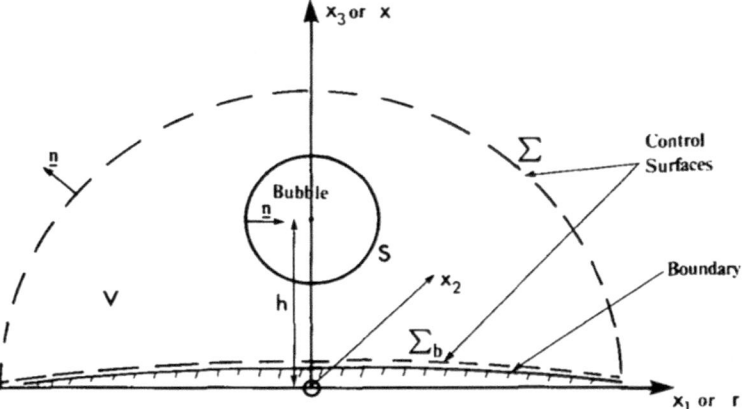

Fig. 4.4 Schematic diagram of selected control body and its boundary. Reprinted with the permission from Ref. [5] Copyright (1982) (Cambridge University Press)

where "f" may be a tensor, scalar, or vector function.

Based on the Reynolds transport theorem and simplifying the second term on the right side of Eq. (4.4), we obtain the following equation [6]

$$\rho\frac{\mathrm{d}}{\mathrm{d}t}\int_{\Sigma\cup\Sigma_b}\Phi n\mathrm{d}A = \rho\frac{\mathrm{d}}{\mathrm{d}t}\int_{V_b}\nabla\Phi\mathrm{d}V = \rho\int_{\Sigma\cup\Sigma_b}\frac{\partial\Phi}{\partial t}\cdot n\mathrm{d}A + \rho\int_{\Sigma\cup\Sigma_b}\frac{\partial\Phi}{\partial n}\nabla\Phi\mathrm{d}A \quad (4.6)$$

Here,

$$\frac{\partial\Phi}{\partial n} = n\cdot\nabla\phi \quad (4.7)$$

where "V_b" represents the volume corresponding to boundary surface "$\Sigma\cup\Sigma_b$".

Based on the non-constant Bernoulli equation, Eq. (4.8) can be obtained as [5]

$$p = p_\infty - \rho\frac{\partial\Phi}{\partial t} - \frac{1}{2}\rho|u|^2 \quad (4.8)$$

where "p_∞" represents the pressure of a fluid at an infinite distance.

Substituting Eq. (4.6) and (4.8) into Eq. (4.4), we obtain

$$F_S = \frac{\mathrm{d}I_S}{\mathrm{d}t} = -\int_{\Sigma\cup\Sigma_b}p_\infty n\mathrm{d}A + \rho\int_{\Sigma\cup\Sigma_b}\left(\frac{1}{2}|\nabla\Phi|^2 n - \frac{\partial\Phi}{\partial n}\nabla\Phi\right)\mathrm{d}A \quad (4.9)$$

where "F_s" represents the rate of change of the impulse of the bubble, that is the combined external force on the bubble.

It can be approximated that the face "S" is coincident with the surface of the face "$\Sigma \cup \Sigma_b$" in the opposite normal direction when the volume of the selected control body is sufficiently small. Thus, the combined external force on the bubble is equal to the integration on the "S" surface is given as [5]

$$\left\{ \int_S + \int_{\Sigma \cup \Sigma_b} \right\} n \mathrm{d}A = 0 \tag{4.10}$$

Substituting Eq. (4.10) into Eq. (4.9), we obtain [5]

$$F_S = F_g + F_\Sigma \tag{4.11}$$

Here,

$$F_g = \int_S p_\infty n \mathrm{d}A = \rho g V e_z \tag{4.12}$$

$$F_\Sigma = -\rho \int_S \left(\frac{1}{2} |\nabla \Phi|^2 n - \frac{\partial \Phi}{\partial n} \nabla \Phi \right) \mathrm{d}A \tag{4.13}$$

where "F_g" represents the buoyancy force on a bubble. "F_Σ" represents the Bjerknes force on a bubble. "e_z" represents the direction vector in the vertical direction.

Based on the relationship between the impulse and the force applied, the Kelvin impulse on the bubble can be expressed as

$$I_S = \int_0^T F_S \mathrm{d}t \tag{4.14}$$

where "T" represents the bubble period.

Based on Lagally theorem, Landweber and Miloh established the following equation [9]

$$\int_S \left(\frac{1}{2} |\nabla \Phi|^2 n - \frac{\partial \Phi}{\partial n} \nabla \Phi \right) \mathrm{d}A = 4\pi \sum_s P_q D_s^q (\nabla \phi')_s \tag{4.15}$$

Here,

$$\Phi = \phi + \phi' \tag{4.16}$$

$$\phi = -P_q D_s^q (1/R) \tag{4.17}$$

$$D_s^q = \frac{\partial^q}{\partial x_s^\alpha \partial x_s^\beta \partial x_s^\gamma} \tag{4.18}$$

where "Φ" represents the velocity potential of the fluid near the bubble in the infinite flow field. "ϕ'" represents the additional velocity potential of the fluid in the vicinity of the bubble produced by the boundary. "q" represents the summation convention number. "$()_s$" represents the summation at the singularity of the velocity potential. "R" represents the instantaneous radius of the bubble.

Substituting Eq. (4.15) into Eq. (4.13), The Bjerknes force can be written as [3]

$$F_\Sigma = -4\pi \sum_s P_q D_s^q (\nabla \phi')_s \tag{4.19}$$

The Bjerknes force can be further simplified as [3]

$$F_\Sigma = -m \nabla \phi'(r_0) \tag{4.20}$$

Integrating the Bjerknes force over time, the Kelvin impulse can be written as [3]

$$I = \int_0^T F_\Sigma \, dt \tag{4.21}$$

4.1.3   Kelvin Impulse of a Spherical Bubble

The Kelvin impulse expression for a spherical bubble is expressed as [10]

$$I = -4\pi \rho \int_0^T R^2 \dot{R} (\nabla \varphi')_{r_0} \, dt. \tag{4.22}$$

"$\dot{R}$" represents the first-order derivative of the instantaneous radius of the bubble with respect to time. "φ" represents the additional velocity potential due to the boundary. "r_0" represents the position of a bubble. "∇" represents the Nabla operator.

To illustrate the spatial properties of the Kelvin pulse, a parameter to describe the sensitivity of the Kelvin impulse direction "θ_k" is defined as [11]

$$H = \lg\left(\sqrt{\left(\frac{\partial \theta_k}{\partial x}\right)^2 + \left(\frac{\partial \theta_k}{\partial y}\right)^2} \right) \tag{4.23}$$

where "H" represents the sensitivity to the Kelvin impulse direction. "θ_k" represents the direction of the Kelvin impulse.

In addition, the point source strength describing the liquid perturbation by the spherical bubble oscillations is expressed as [10]

$$m = 4\pi R^2 \dot{R} \tag{4.24}$$

where "m" represents the bubble point source strength.

4.1.4 Kelvin Impulse of a Cylindrical Bubble

The Kelvin impulse expression for a cylindrical bubble is expressed as [4]

$$I = -4\pi\rho h \int_{0}^{T} \left(\nabla\varphi'\right)_{z_0} R\dot{R}dt \tag{4.25}$$

where "h" represents the axial height of the cylindrical bubble. "z_0" represents the center of the cylindrical bubble.

Furthermore, the point source strength describing the liquid perturbation by the oscillation of the cylindrical bubble is expressed as [4]

$$m = 2\pi h R\dot{R} \tag{4.26}$$

4.2 Boundary Processing Method

4.2.1 Weiss Theorem

Weiss theorem is a mathematical method for treating the boundary of a spherical particle near a bubble. This theorem allows the superposition of an image bubble and a uniformly distributed linear sink to characterize the effects of the particle on the flow field and the bubble. Figure 4.5 shows the schematic diagram of the boundary treatment of a single spherical particle based on Weiss theorem [12]. In Fig. 4.2, the center of the particle is set as the origin of the coordinates. The particle with radius R_p and the cavitation bubble with radius R are represented by circles in grey and white, respectively. The blue dashed circle and the red dashed line represent the image bubble and the linear sink, respectively. The initial positions of the bubble and the image bubble are represented by r_0 and r_i, respectively, and both of them are located on the x-axis. The details of the coordinate positions of the bubble, the

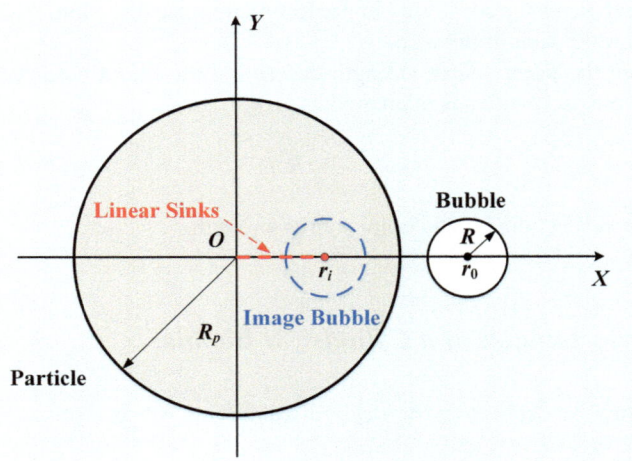

Fig. 4.5 Schematic diagram of the boundary treatment of a single spherical particle based on Weiss theorem [12]. Reprinted with the permission from Ref. [10] Copyright (2022) (Elsevier)

Table 4.1 Detailed information on the boundary treatment of a single particle

Objects	Positions	Intensity/density	Velocity potential
Bubble	$r_0 = (l, 0)$	m	$\varphi_0 = -\frac{m}{4\pi}\frac{1}{|r-r_0|}$
Image bubble	$r_i = \left(\frac{R_p^2}{l},\ 0\right)$	$\frac{mR_p}{l}$	$\varphi_{\text{IB}} = -\frac{mR_p}{4\pi l}\frac{1}{|r-r_i|}$
Linear sink	From O to r_i	$-\frac{m}{R_p}$	$\varphi_{\text{LS}} = \frac{m}{4\pi R_p}\int\limits_{0}^{R_p^2/l}\frac{ds}{|r-(s,0)|}$

image bubble and the linear sink, as well as the associated velocity potentials, are listed in Table 4.1.

4.2.2 Image Method

Taking the wall-particle-bubble model as an example, this section introduces the basic principle of the image method. The effects of the particle and wall on the bubble dynamics are evaluated by introducing multiple image bubbles and linear sinks. Figure 4.6 shows the boundary treatment diagram of the bubble between the particle and the rigid wall based on the image method. The interaction of the bubble, particle and rigid wall surface with the surrounding liquid can be represented as a sum of three image bubbles and two linear sinks.

Fig. 4.6 Schematic diagram of the particle–wall boundary treatment based on the image method. Reprinted with the permission from Ref. [13] Copyright (2024) (Springer Singapore)

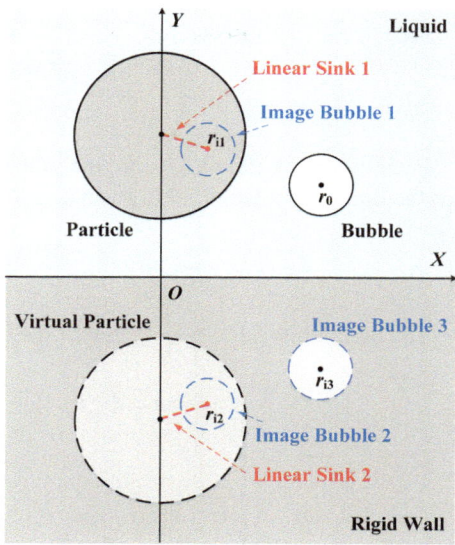

4.2.3 Conformal Transformation

This section treats a wall with an arched bulge through a conformal transformation method. By transforming the complex boundary into an easily analyzable planar boundary, the additional distribution of the liquid velocity potential induced by the boundary is then calculated. Firstly, the wall with an arched cylinder bulge is transformed into an angular wall consisting of two walls by a fractional linear transformation. Next, the angular wall is further transformed into a flat wall by a power function transformation. Finally, the origin and axes on the walls are corrected again with the fractional linear transformation to ensure that the transformed walls are correctly represented on the auxiliary plane. In the auxiliary plane, the principle of the image method is used to represent the role of the wall by the virtual cavitation bubble, and then the flow field distribution functions such as the complex potential of the liquid and the complex velocity are calculated in the auxiliary plane. The flow field distribution function in the auxiliary plane is converted to the original physical model by performing a conformal inverse transformation of the spatial coordinates in the function. From this, it is obtained the additional liquid velocity potential function generated by the convex wall. The calculated velocity potential information is imported into the Kelvin impulse theory model to achieve a theoretical calculation of the liquid flow field distribution. Through the above process of conformal transformation, it can be obtained that the coordinate functions in the auxiliary plane and in the plane to be analyzed can be expressed as

$$w = -\frac{h_p}{2} \cdot \frac{\left(\frac{z-h_p/2}{z+h_p/2}\right)^{\pi/(\pi-\theta_p)} + 1}{\left(\frac{z-h_p/2}{z+h_p/2}\right)^{\pi/(\pi-\theta_p)} - 1} \tag{4.27}$$

By introducing Eq. (4.27) into the potential function of the reset, the expression for the liquid reset potential is given as

$$F(z, t) = \frac{m}{2\pi h} \ln(w(z) - w_0(z)) + \frac{m}{2\pi h} \ln(w(z) - \overline{w_0}(z)) \tag{4.28}$$

The additional velocity potential of the liquid is obtained as follows:

$$\varphi' = \text{real}\left(F(z, t) - \frac{m}{2\pi h} \ln(z - z_0)\right) \tag{4.29}$$

4.3 Typical Kelvin Impulse Results

4.3.1 Bubble Near a Spherical Particle

This section discusses the bubble dynamics near a spherical particle. Figure 4.7 shows the velocity distribution of the liquid with different distances between the particle and the bubble. It can be noticed that the flow velocity of the liquid to the right of the particle is lower and there exists a low velocity region. Comparing Fig. 4.7a–c, as the bubble moves away from the particle, the influence of the particle on the flow of the liquid decreases, and the difference in the liquid flow velocity between the left and right sides of the bubble gradually decreases.

Figure 4.8 shows the spatial distribution of the Kelvin impulse near the particle–wall surface. The grey circle and the white area at the bottom refer to the particle and the wall, respectively. In Fig. 4.8, as the bubble is closer to the particle, the Kelvin impulse is relatively strong and directed approximately towards the center of the particle. As the bubble is closer to the wall, the Kelvin impulse is relatively strong and direction approximately perpendicular to the wall. When the bubble is in other positions influenced by the wall and the particle, the Kelvin impulse direction has a complex spatial distribution, which is no longer perpendicular to the wall or towards the center of the particle. Furthermore, a zone of weak Kelvin impulse is observed at the center between the particle and the wall. This region displays a high spatial sensitivity of the Kelvin impulse direction and is significantly influenced by the position of the bubble.

Figure 4.9 shows the comparison between theoretical and experimental predictions of the average speed of the bubble centroid movement in during the first period of bubble oscillation. The black dots represent the results of the experiment. The red

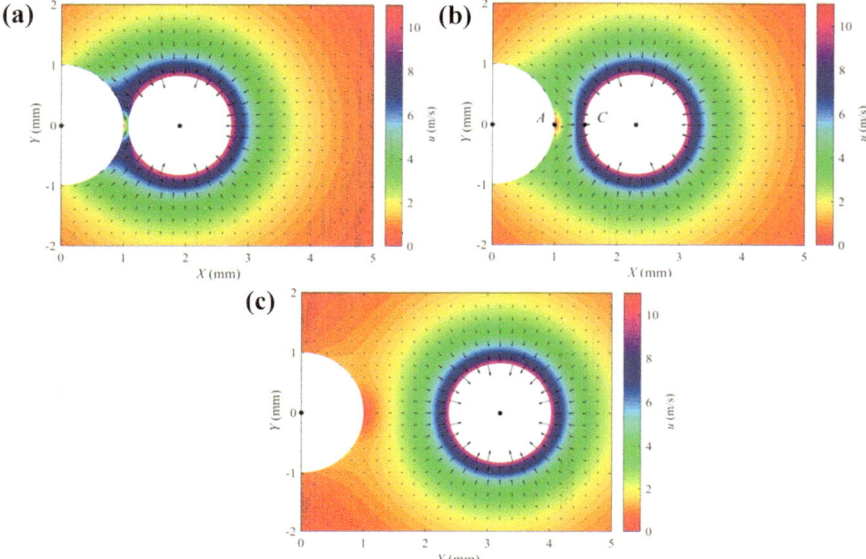

Fig. 4.7 Velocity distribution of the liquid with different distances between the particle and the bubble. Reprinted with the permission from Ref. [10] Copyright (2022) (Elsevier)

Fig. 4.8 Spatial distribution of the Kelvin impulse near the particle–wall surface

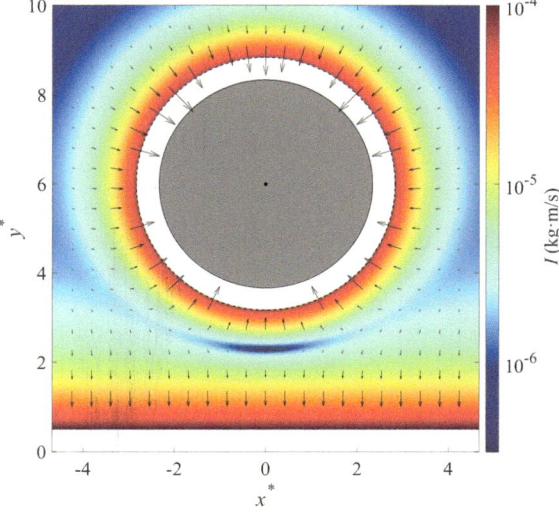

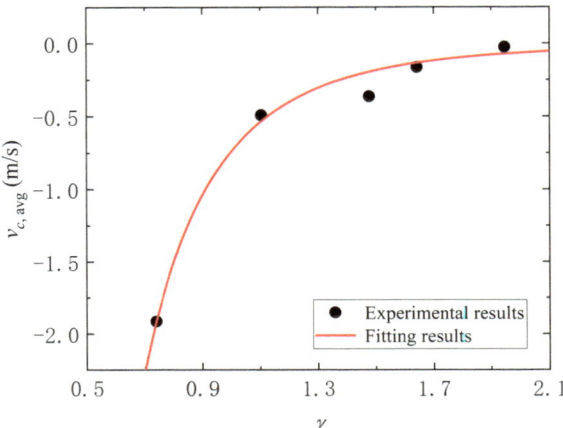

Fig. 4.9 Comparison between theoretical and experimental predictions of the average speed of movement of the bubble centroid during the first collapse of the bubble. Reprinted with the permission from Ref. [10] Copyright (2022) (Elsevier)

line represents the fitted curve calculated. From the figure, it can be seen that the average velocity of the bubble centroid movement gradually increases and tends to zero with increasing distance between the bubble and the particle, and the growth rate gradually decreases. Notably, there are spatial limitations to the method of fitting average velocity based on Kelvin impulse. In the case where the bubble is very close to the particle, the theoretical model cannot accurately predict bubble motion behaviors.

$$v_{c,\text{avg}} = \frac{1}{\alpha} I \tag{4.30}$$

$$I = -\frac{8\sqrt{6}}{9}\pi((p_0 - p_v)\rho)^{\frac{1}{2}} R_{\text{max}}^5 B\left(\frac{7}{6}, \frac{3}{2}\right) \frac{R_p^3}{l^2 \left(l^2 - R_p^2\right)^2} \tag{4.31}$$

Figure 4.10 shows the variation of the Kelvin impulse intensity with dimensionless bubble positions under different radius ratios and dimensionless spacings of the particle pair. Figure 4.10a and b show the effect of radius ratio and distance between the upper particle and lower particle, respectively. In Fig. 4.10a, as the particle size difference increases, the particle sensitivity to the position of the bubble increases, and the Kelvin impulse equilibrium position is tilted towards the smaller particle below. In Fig. 4.10b, as the particle pair spacing increases, the sensitivity of the impulse strength to the position of the bubble gradually decreases, and the Kelvin impulse equilibrium position is gradually tilted towards the lower particle.

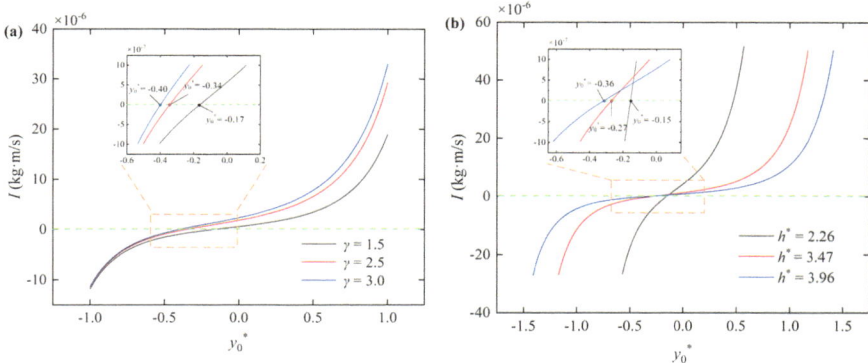

Fig. 4.10 Variation of the Kelvin impulse intensity with dimensionless bubble positions under different radius ratios and dimensionless spacings of the particle pair. Reprinted with the permission from Ref. [14] Copyright (2024) (American Institute of Physics)

4.3.2   Bubble Near a Wall with a Bulge

Figure 4.11 shows the dominant region of the Kelvin impulse distribution near a wall with a bulge. Figure 4.11a shows the dominant region distributions of the Kelvin impulse components relative to the bulge radius and the bulge-bubble distances. Figure 4.11b shows the dominant region distributions of the Kelvin impulse components relative to the azimuth angle of the bubble and the bulge-bubble distances. In the figure, the shaded area represents the non-physical case, where the bubbles are primed inside the bulge. Without considering the above regions, it can be divided into three regions dominated by the flat wall (blue region), the hemispherical bulge (red region) and both of them (white region) for the Kelvin impulse, respectively. It can also be noticed that the region dominated by the hemispherical bulge (red region) is narrower in Fig. 4.11b. And when the distance between the bubble and the bulge increases to a certain extent, the Kelvin impulse of the dominant bubble is completely dominated by the flat-wall (blue region).

Figure 4.12 shows the spatial distributions of the Kelvin impulse near a wall with a bulge. As shown in the figure, the Kelvin impulse intensity near the bulge can be distinguished into three regions, where Region 2 receives the strongest influence from the bulge and the wall, and the Kelvin impulse intensity is greater than Region 1 and Region 3.

Figure 4.13 shows the spatial distribution of the jet attraction zone near the bulge. As the bubble moves horizontally to the right away from the bulge, the area of the jet attraction zone in the vertical direction initially increases and then gradually decreases, and eventually becomes steady. Therefore, there is a maximum jet attraction area in the vertical direction.

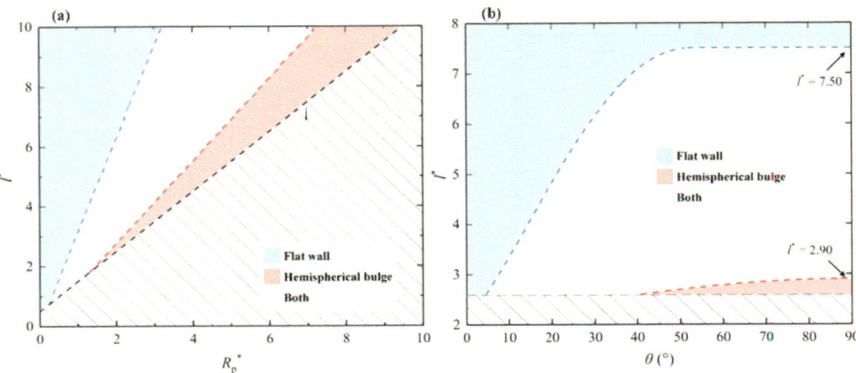

Fig. 4.11 Dominant zone distributions of the Kelvin impulse near a wall with a bulge. Reprinted with the permission from Ref. [15] Copyright (2024) (American Institute of Physics)

Fig. 4.12 Spatial distributions of the Kelvin impulse near a wall with a bulge. Reprinted with the permission from Ref. [15] Copyright (2024) (American Institute of Physics)

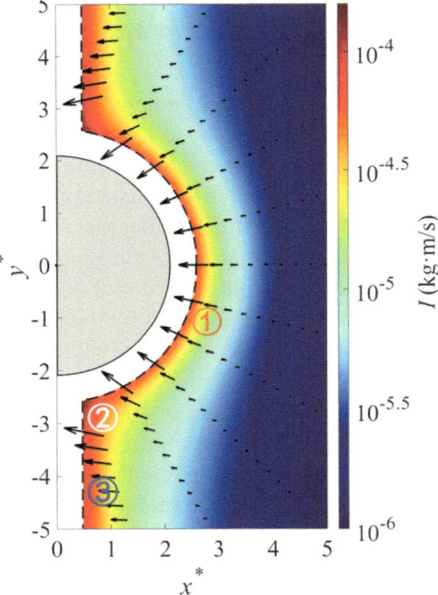

4.3.3 Bubble Within a Spherical Droplet

This section discusses the Kelvin impulse distribution characteristics of a bubble within spherical a droplet. Figure 4.14 shows the liquid velocity field inside a spherical droplet for different bubble-droplet eccentricities. Figure 4.14a–d represent the cases of small radius ratio, medium radius ratio, large radius ratio and very large

Fig. 4.13 Spatial distribution of the jet attraction zone around the bulge. Reprinted with the permission from Ref. [15] Copyright (2024) (American Institute of Physics)

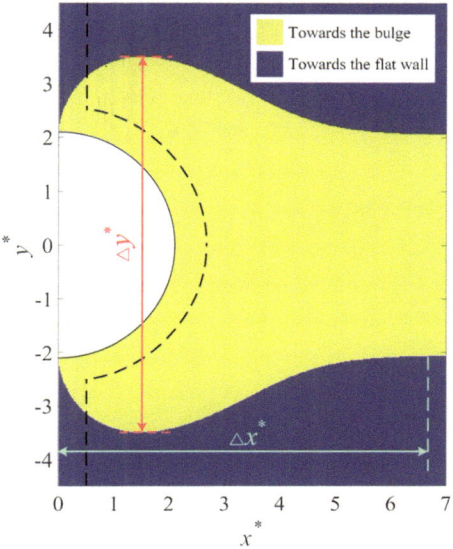

radius ratio, respectively. As shown in the figure, when the radius ratio is relatively small (Fig. 4.14a), the liquid velocity around the bubble wall essentially shows an isotropic distribution, which is relatively unaffected by the surface of the droplet. As the radius ratio increases (Fig. 4.14b–d), the perturbation of the liquid inside the droplet by the bubble collapse becomes more intense and the anisotropy of the liquid velocity distributions around the bubble wall and the droplet surface gradually increases. Specifically, when the radius is relatively large (Fig. 4.14d), the contraction velocity of the right side of the bubble wall is significantly stronger and stronger than the other sides of the bubble, and in addition, the movement velocity of the right surface of the droplet is similarly significantly stronger and stronger than the other sides.

Figure 4.15 shows the distribution of the surface velocity of the bubble and droplet with different eccentricities. Figure 4.15a and b show the distribution patterns of bubble wall and droplet surface velocities, respectively. As shown in Fig. 4.15a, when the bubble and droplet are located at concentric positions, the liquid velocity around the bubble wall is uniformly isotropic. When the bubble is located in an eccentric position, the distribution of the velocity around the bubble wall is no longer uniform, with the lowest velocity on the left vertex of the bubble wall and the highest velocity on the right vertex. As the eccentricity increases, the difference in the liquid velocity distribution around the bubble wall gradually becomes larger, where the minimum velocity decreases and the maximum velocity increases. In Fig. 4.15b, the velocity distribution pattern on the droplet surface is consistent with Fig. 4.15a, with the minimum velocity decreasing and the maximum velocity increasing with increasing eccentricity.

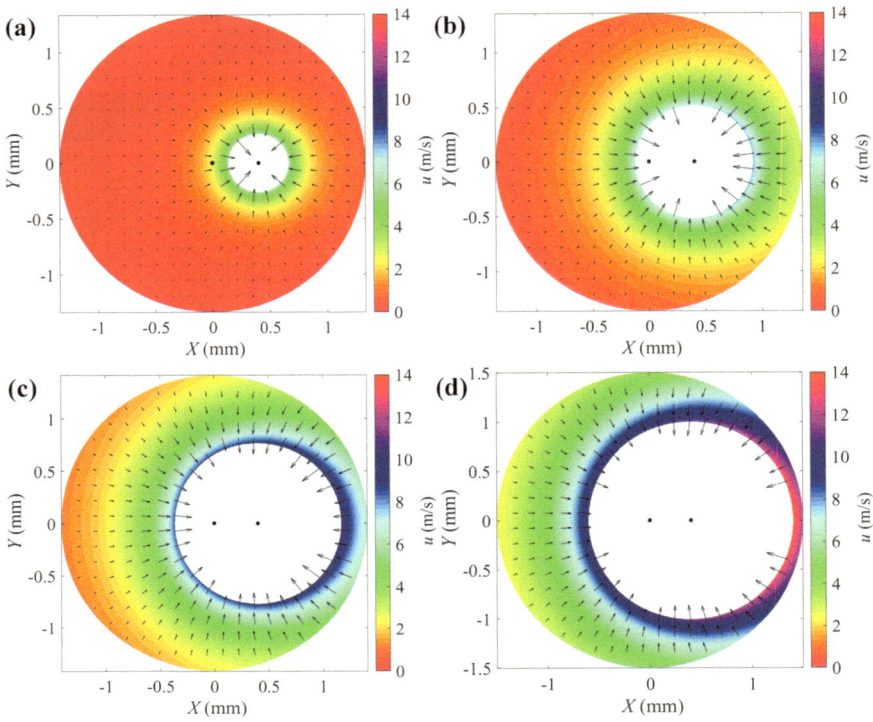

Fig. 4.14 Liquid velocity field within a spherical droplet with different bubble-droplet eccentricities

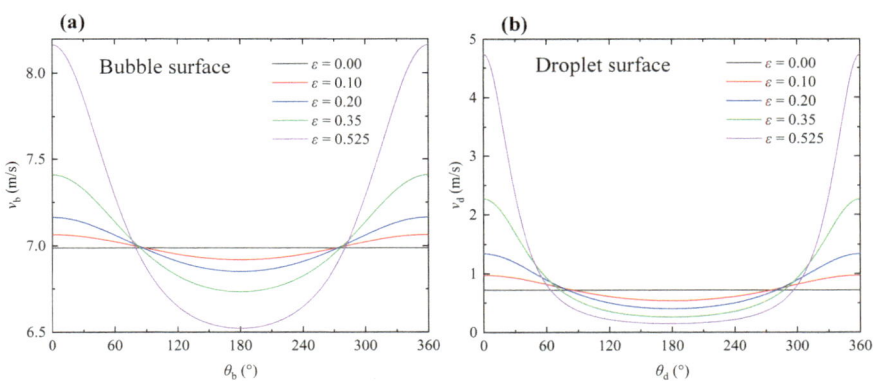

Fig. 4.15 Distribution of the surface velocity of the bubble and droplet with different eccentricities

4.3.4 Bubble Near a Hydrofoil

This section discusses the Kelvin impulse distribution characteristics of a bubble near a hydrofoil. Figure 4.16 shows the liquid velocity distribution when the bubble is located near different positions of the hydrofoil. As shown in the figure, there is a region of lower velocities in the bubble region near the hydrofoil, while the area of the bubble wall away from the hydrofoil shows higher velocities. This characteristic distribution of the velocity field leads to the formation of distinct depressions on the side of the bubble wall away from the hydrofoil. The shape and direction of such depressions vary with the azimuthal angle of the bubble.

Figure 4.17 shows the distribution of Kelvin impulse around the elliptical wall with different short-to-long axis ratios. In Fig. 4.17a–d, as the short-to-long axis ratio gradually increases, the shape becomes closer to a circle, and the curvature difference and the Kelvin impulse intensity distribution decrease. However, the influence of

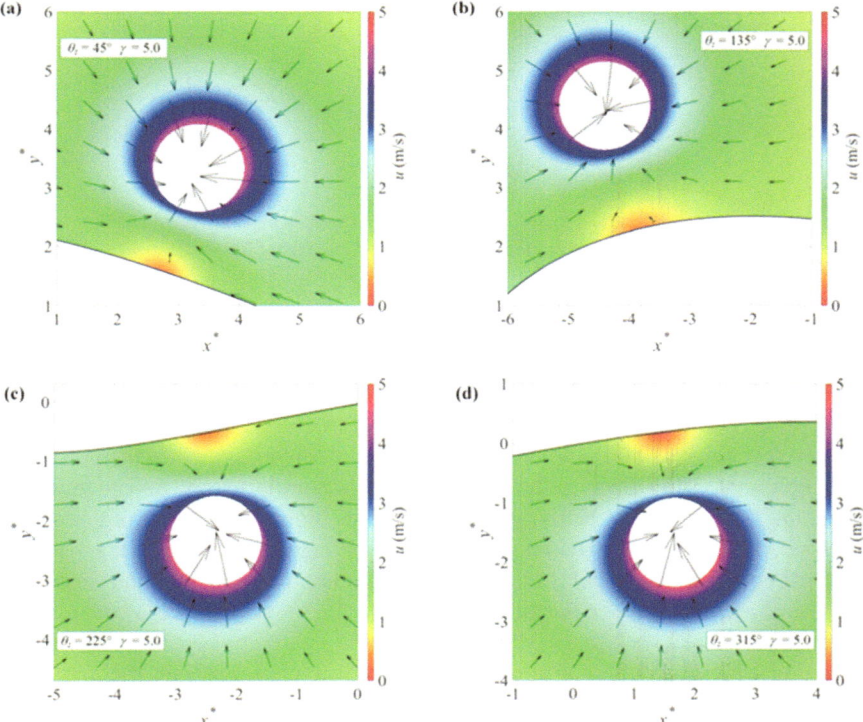

Fig. 4.16 Liquid velocity distribution when the bubble is located near different positions of the hydrofoil. Reprinted with the permission from Ref. [11] Copyright (2024) (American Institute of Physics)

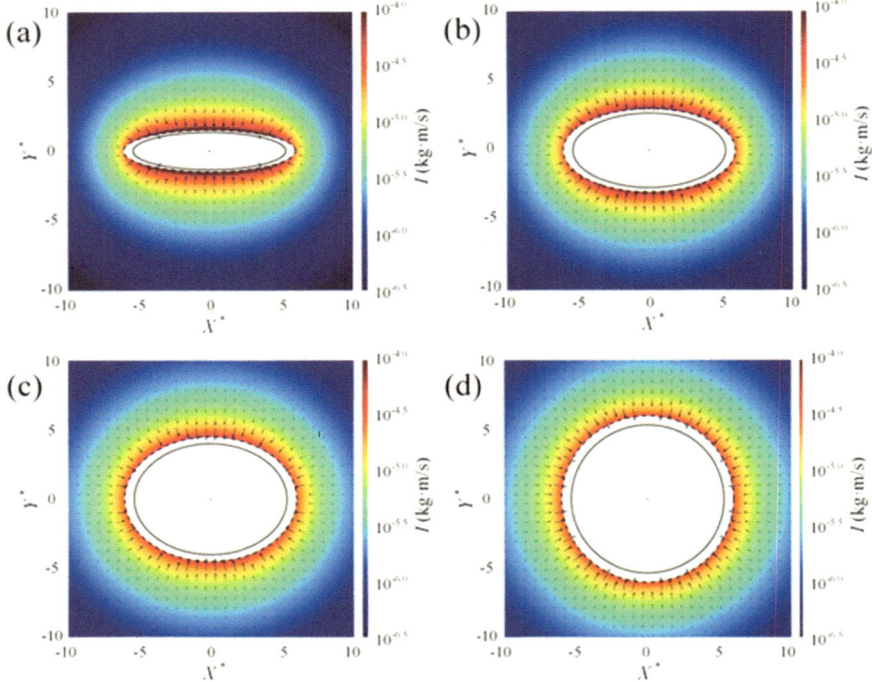

Fig. 4.17 The distribution of Kelvin impulse around the elliptical wall with different short-to-long axis ratios. Reprinted with the permission from Ref. [16] Copyright (2024) (Elsevier)

the elliptic wall on the surrounding spatial region becomes more extensive as the short-to-long axis ratios increase.

Figure 4.18 shows the Kelvin impulse sensitivity distribution of the bubble near a hydrofoil. The sensitivity index can be expressed as Eq. (4.23) [17]. The blank area between the hydrofoil and the dashed line in the figure indicates the portion of the bubble that is too close to the hydrofoil and is not discussed. As shown in the figure, the Kelvin pulse sensitivity of the bubble is highest in the tail region of the hydrofoil, followed by the head region. The middle region has the lowest sensitivity due to its concave shape. This indicates that the Kelvin pulse variation is very small and almost negligible when the bubble is close to the lower part of the center of the hydrofoil.

Figure 4.19 shows the categorization of the difference between the azimuthal angle of the bubble and the direction angle of the Kelvin impulse near an elliptical wall in the first quadrant. $\Delta\theta$ represents the difference between the azimuthal angle of the bubble and the direction angle of the Kelvin impulse. The blue, green and orange regions represent $\Delta\theta < 5°$, $5° < \Delta\theta < 15°$ and $\Delta\theta > 15°$, respectively. As shown in the figure, in the first quadrant, the elliptical wall has a very limited effect

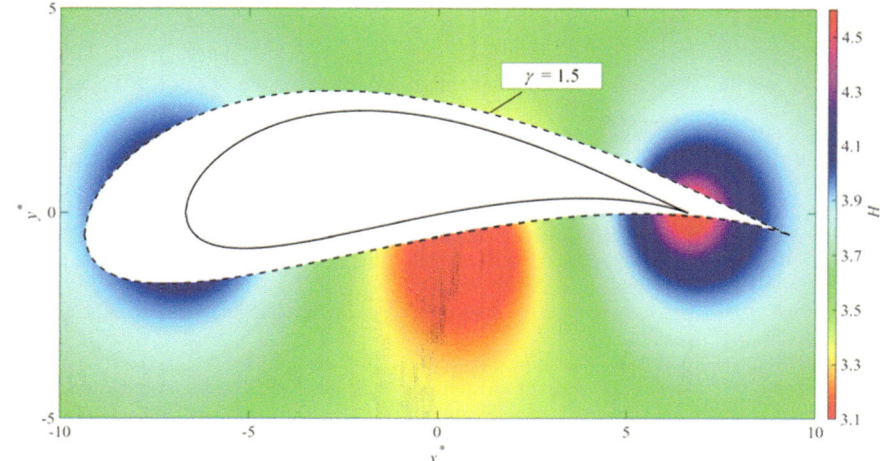

Fig. 4.18 Kelvin impulse sensitivity distribution of the bubble near a hydrofoil. Reprinted with the permission from Ref. [11] Copyright (2024) (American Institute of Physics)

Fig. 4.19 Categorization of difference between the azimuthal angle of the bubble and the direction angle of the Kelvin impulse near an elliptical wall in the first quadrant. Reprinted with the permission from Ref. [16] Copyright (2024) (Elsevier)

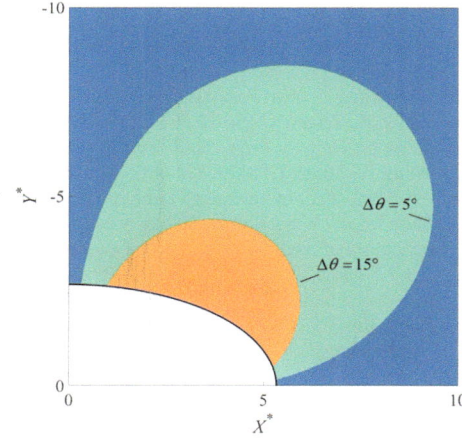

on the nearby Kelvin impulse direction when the bubble is located in the horizontal and vertical directions. In contrast, within the other regions of the first quadrant, the effect of the elliptical walls is significant. In the region of $\Delta\theta < 5°$, the elliptical wall has a similar effect to the circular wall.

References

1. Plesset MS, Prosperetti A (1977) Bubble dynamics and cavitation. Annu Rev Fluid Mech 9:145–185
2. Best JP (1991) The dynamics of underwater explosions. University of Wollongong
3. Best JP, Blake JR (1994) An estimate of the Kelvin impulse of a transient cavity. J Fluid Mech 261:75–93
4. Wang X, Wu G, Shen J et al (2023) Research on the collapse dynamics of a restricted cavitation bubble near a right-angle wall based on Kelvin impulse theory. Phys Fluids 35(7):073335
5. Blake JR, Cerone P (1982) A note on the impulse due to a vapour bubble near a boundary. ANZIAM J 23(4):383–393
6. Blake JR (1988) The Kelvin impulse: application to cavitation bubble dynamics. ANZIAM J 30(2):127–146
7. Reynolds O (1903) The sub-mechanics of the universe. University Press
8. Leal LG (2007) Advanced transport phenomena: fluid mechanics and convective transport processes. Cambridge University Press, Cambridge
9. Landweber L, Miloh T (1980) Unsteady Lagally theorem for multipoles and deformable bodies. J Fluid Mech 96(1):33–46
10. Wang X, Wu G, Zheng X et al (2022) Theoretical investigation and experimental support for the cavitation bubble dynamics near a spherical particle based on Weiss theorem and Kelvin impulse. Ultrason Sonochem 89:106130
11. Shen J, Li S, Wang X et al (2024) Theoretical and experimental investigation of a bubble collapsing near an asymmetric hydrofoil. Phys Fluids 36(2):587
12. Sun X, Xia G, You W et al (2023) Effect of the arrangement of cavitation generation unit on the performance of an advanced rotational hydrodynamic cavitation reactor. Ultrason Sonochem 99:106544
13. Yu J, Wang X, Shen J et al (2024) Physics of cavitation near particles. J Hydrodyn 36(1):102–118
14. Wang X, Zhang C, Su H et al (2024) Research on cavitation bubble behaviors between a dual-particle pair. Phys Fluids 36(2):78–94
15. Wang X, Zhang C, Shen J et al (2024) Influence of a hemispherical bulge on a flat wall upon the collapse jet of cavitation bubbles. Phys Fluids 36(3):3236
16. Shen J, Li S, Wang C et al (2024) Investigation on the effects of an elliptical wall on the dynamic behaviors of a bubble restricted by two parallel plates. Ultrason Sonochem 107:106915
17. Shen J, Liu Y, Wang X et al (2023) Research on the dynamics of a restricted cavitation bubble near a symmetric Joukowsky hydrofoil. Phys Fluids 35(7):546
18. Benjamin TB, Ellis AT (1966) The collapse of cavitation bubbles and the pressures thereby produced against solid boundaries. Philos Trans Royal Soc Lond Ser A Math Phys Sci 12:221–240
19. Blake JR, Taib BB, Doherty G (1986) Transient cavities near boundaries. Part 1. Rigid boundary. J Fluid Mech 170:479–497
20. Blake JR, Taib BB, Doherty G (1987) Transient cavities near boundaries Part 2. Free surface. J Fluid Mech 181:197–212
21. Blake JR, Gibson DC (1987) Cavitation bubbles near boundaries. Annu Rev Fluid Mech 19(1):99–123
22. Brujan EA, Pearson A, Blake JR (2005) Pulsating, buoyant bubbles close to a rigid boundary and near the null final Kelvin impulse state. Int J Multiphase Flow 31(3):302–317
23. Supponen O, Obreschkow D, Tinguely M et al (2016) Scaling laws for jets of single cavitation bubbles. J Fluid Mech 802:263–293
24. Andrews ED, Peters IR (2022) Modeling bubble collapse anisotropy in complex geometries. Phys Rev Fluids 7(12):123601
25. Wang Q, Mahmud M, Cui J et al (2020) Numerical investigation of bubble dynamics at a corner. Phys Fluids 32(5):053306
26. Li S, Liu Y, Wang Q et al (2021) Dynamics of a buoyant pulsating bubble near two crossed walls. Phys Fluids 33(7):5336

27. Li S, Zhang A, Cui P et al (2023) Vertically neutral collapse of a pulsating bubble at the corner of a free surface and a rigid wall. J Fluid Mech 962:A28
28. Wang X, Li S, Shen J et al (2024) Dynamic behaviors of a bubble near a rectangular wall with a bulge. Phys Fluids 36(2):4985
29. Xu P, Li B, Ren Z et al (2023) Dynamics of a laser-induced buoyant bubble near a vertical rigid boundary. Phys Rev Fluids 8(8):083601
30. Ren Z, Han H, Zeng H et al (2023) Interactions of a collapsing laser-induced cavitation bubble with a hemispherical droplet attached to a rigid boundary. J Fluid Mech 976:A11

Chapter 5
Visualization of Bubble Collapse Dynamics

This chapter focuses on the visualization of bubble collapse based on high-speed photography. Firstly, the high-speed camera platform and related devices are introduced, and the layout of the experimental platform is described in detail with several examples. Secondly, typical physical behaviors such as the morphological evolution process of the bubble, the jet, and the shock wave are analyzed. Thirdly, the dynamics of the bubble inside the droplet are also analyzed.

5.1 High-Speed Photography Experimental Platform

5.1.1 Experimental System

This section presents the key equipments of the experimental platform for high-speed photography of the laser-induced bubble, including the high-speed camera, laser generator, time-delay generator and others. Figure 5.1 shows the schematic of the high-speed camera platform for the laser-induced bubble, respectively. It can be observed that the experimental system mainly consists of the high-speed camera, the laser generator, the time-delay generator, the computer, the focusing lens, the three-dimensional translation platform, the light source, and other equipment. In particular, the high-speed camera is employed to capture the complete dynamical behaviors of the bubble, such as the morphological evolution and the collapsing jet during the bubble growth and collapse process. The laser generator is employed to generate the pulsed laser light to provide energy conditions for the generation of the bubble. The delay generator is employed to coordinate the laser generator with the high-speed camera. A focusing lens can be employed to adjust the position of the laser to ensure that the bubble is generated at a fixed position. The three-dimensional translation platform can control the relative position of the bubble and the experimental material. The light source can supplement light for the photography process. Table 5.1 shows

© The Author(s), under exclusive license to Springer Nature Switzerland AG 2024

X. Wang et al., *Fundamentals of Single Cavitation Bubble Dynamics*,
SpringerBriefs in Energy, https://doi.org/10.1007/978-3-031-75041-0_5

the main experimental devices and parameters in the high-speed photography system. It mainly includes the actual photo and key parameters.

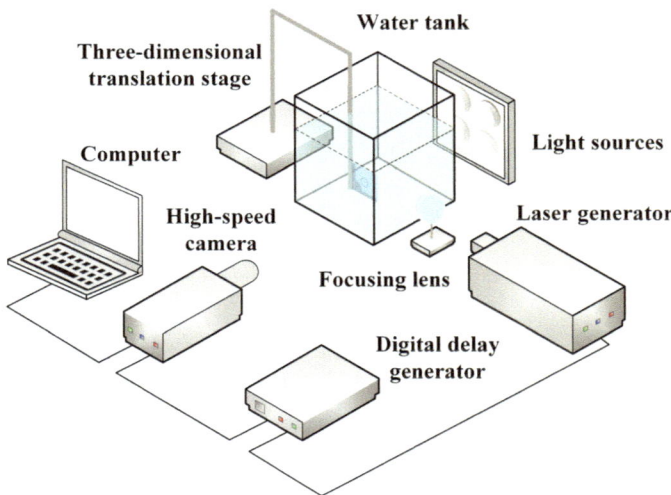

Fig. 5.1 The schematic and physical figures of the high-speed camera platform for the laser-induced bubble. Reprinted with the permission from Ref. [1] Copyright (2024) (American Institute of Physics)

Table 5.1 The main experimental devices and parameters in the high-speed photography system

Device	Model/material	Parameters
High-speed camera	Qianyanlang X113 series	Max frame rate: 384,000 fps Current frame rate: 47,619 fps
Digital delay generator	Stanford research systems DG535	Delay resolution: 5 ps Channel-to-channel jitter: 50 ps
Laser generator	Anshan ZY laser technology Co., Ltd. Penny-100A-SC	Max pulse energy: 100 mJ Wavelength: 532 nm
Focusing lens	Thorlabs LMH-10X-532	Magnification: $10\times$
Three-dimensional translation stage	Liansheng optics LSDP-50JS	Adjustable range: 100 mm $\times$ 100 mm $\times$ 100 mm
Water tank	Transparent acrylic plate	Size: 150 mm $\times$ 150 mm $\times$ 150 mm

Reprinted with the permission from Ref. [1] Copyright (2024) (American Institute of Physics)

5.1.2   Operational Process

This section describes the experimental procedure of the high-speed photography system. Before the experiment, the required boundary materials are processed and produced by 3D printing or laser cutting. In the preparation stage of the experiment, the laser generator is started for pre-combustion, and parameters such as trigger mode and working voltage are adjusted at the same time. Subsequently, the high-speed camera is started and connected to the computer, while adjusting the shooting frame rate, shutter speed, white balance and so on. In addition, turn on the light source, adjust the position of the light source and arrange the glass plate to make the camera screen brightness uniform, and adjust the brightness of the light source to make it match the shooting frame rate and shutter speed of the high-speed camera to ensure that the brightness of the screen is moderate. Then, adjust the position of the high-speed camera so that the laser focusing position is located near the center of the camera screen, and adjust the focal length of the camera lens so that it focuses on the initial position of the bubble, and determine the scale of the screen. Then adjust the delay parameters of each channel of the delay generator to achieve the synchronous triggering of the high-speed camera and the laser generator, and to ensure that the dynamics of the bubble occur during the work process of the camera. Next, the energy of the laser generator is adjusted to control the size of the bubble, and the relative positions of the experimental material and the bubble are controlled by moving the 3D translation stage, so as to achieve the parametric experiment.

5.1.3   Experimental Layout

This section describes the layout of the high-speed photography experimental platform, including a clamping device for holding the boundary material and performing three-dimensional translation, as well as the equipment to generate and control the bubbles. Figure 5.2 shows the experimentally relevant arrangement of a bubble near particles in a semi-infinite liquid environment. Figure 5.2a shows a physical representation of the boundary material fixation and translation device consisting of a grasping clamp with a three-dimensional translation platform. Figure 5.2b and c show physical objects of spherical particles and a wall with hemispherical bulges, respectively. Here, for the sake of convenience, the particles are attached by glue to a needle, which is attached to a fixing plate and fixed to a gripping clamp. The wall with a hemispherical projection is fixed directly to the gripping clamp, thus enabling three-dimensional translational control.

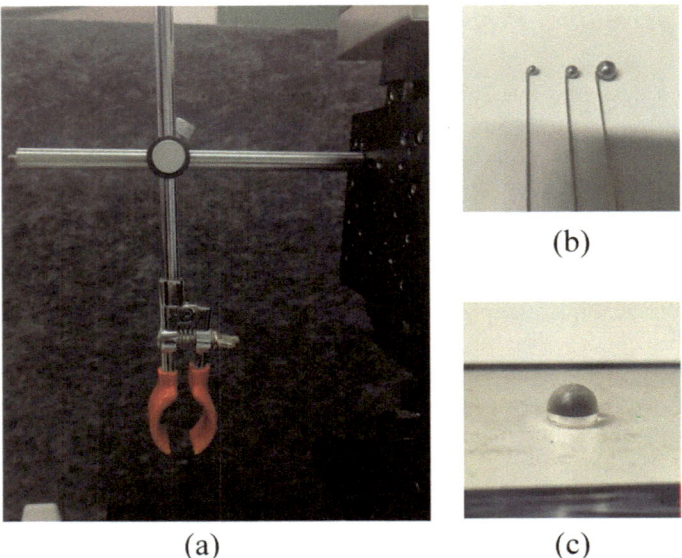

(a)   (c)

Fig. 5.2 Experimentally relevant arrangement of a bubble near particles in a semi-infinite liquid environment

Figure 5.3 shows a schematic and an actual diagram of the experimental arrangement of the bubble in a dangling droplet. The system consists of a syringe pump, an infusion tube, a flat-tipped needle, and a three-dimensional translation platform. In the experiment, a stable droplet can be obtained at the needle by precisely controlling the start-stop time and flow rate of the syringe pump to make the syringe advance at a uniform speed. The dangling droplet presents an ellipsoidal shape due to the gravity. The laser-focusing position can be aligned with the droplet position by adjusting the position of the 3D translation platform.

Figure 5.4 shows a schematic and an actual diagram of the experimental arrangement of the bubble in an attached droplet. The system consists of a syringe pump, an infusion tube, a flat-tipped needle and a glass plate with a hydrophobic surface. In this experimental system, the flat-tipped needle is inserted and fixed in a cylindrical penetration hole in the glass plate. By controlling the start/stop time and flow rate of the syringe pump, a droplet of controlled position and size is formed in the cylindrical groove of the hydrophobic glass plate. The droplet presents a quasi-ellipsoidal shape because of the hydrophobicity of the glass plate surface and gravity. Finally, the location of the droplet is moved to the laser-focusing position by fixing the system on a 3D translation platform.

Fig. 5.3 Schematic and actual diagram of the experimental arrangement of the bubble in a dangling droplet

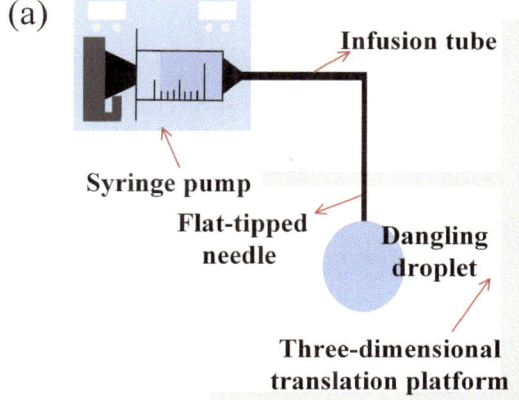

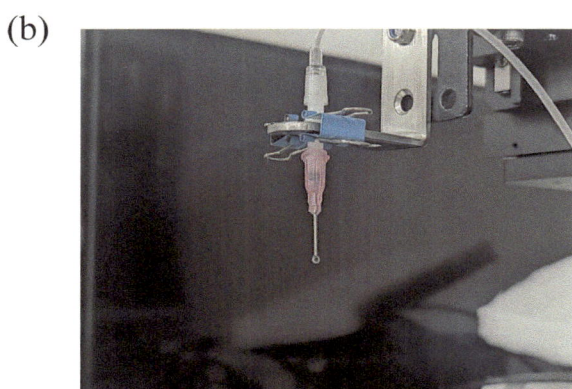

5.2 Bubble Morphology Evolution

5.2.1 Complete Bubble Dynamics Process

Figure 5.5 shows the complete dynamics of a bubble near a particle through. Figure 5.5a–c show the images of the bubble during three periods, respectively. As shown in Fig. 5.5a, the first period of the bubble can be divided into the growth stage and the collapse stage. During the growth stage, the bubble grows to the maximum size and usually maintains a spherical shape. During the collapse stage, the bubble wall on the left side of the bubble moves faster, the bubble deforms and collapses to its minimum volume. As shown in Fig. 5.5b, the second period of the bubble can be divided into the rebound stage and the second collapse stage. The bubble is in contact with the particle surface at this period and presents a more complicated shape. As

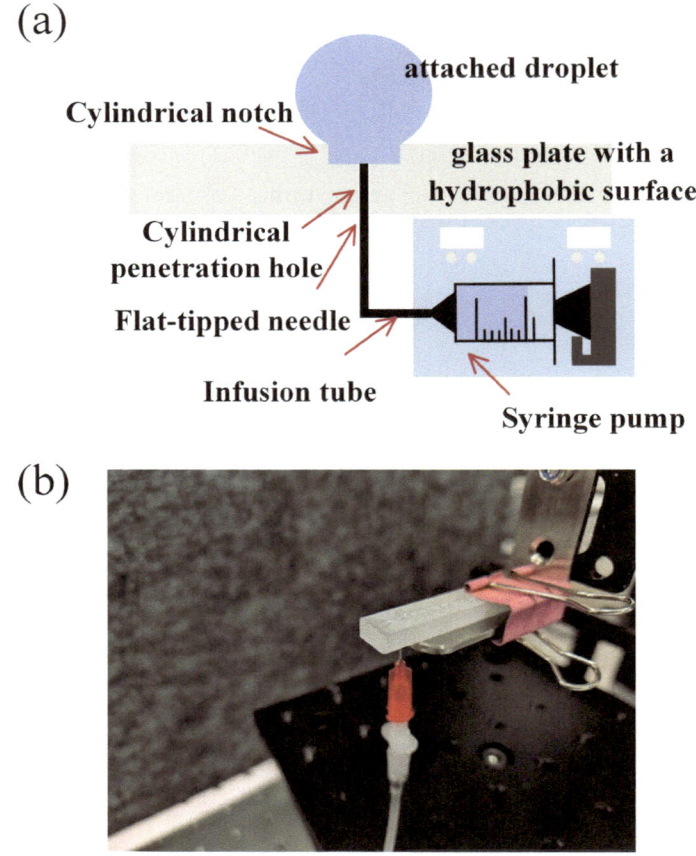

Fig. 5.4 Schematic and actual diagram of the experimental arrangement of the bubble in an attached droplet

shown in Fig. 5.5c, in the third period of the bubble, the bubble size is smaller and less visible, and the motion behaviors are more irregular.

5.2.2 Morphology Evolution in the First Period

Figure 5.6 shows several typical morphology evolutions of the bubble during its first period for different distances between the bubble and the particle. Figure 5.6a–c represent three typical cases from close to far from the particle, respectively. When the bubble is close to the particle, the bubble presents a mushroom shape and forms a jet towards the particle at the end of the collapse process. When the bubble is at a medium distance from the particle, the bubble presents a pear shape during the

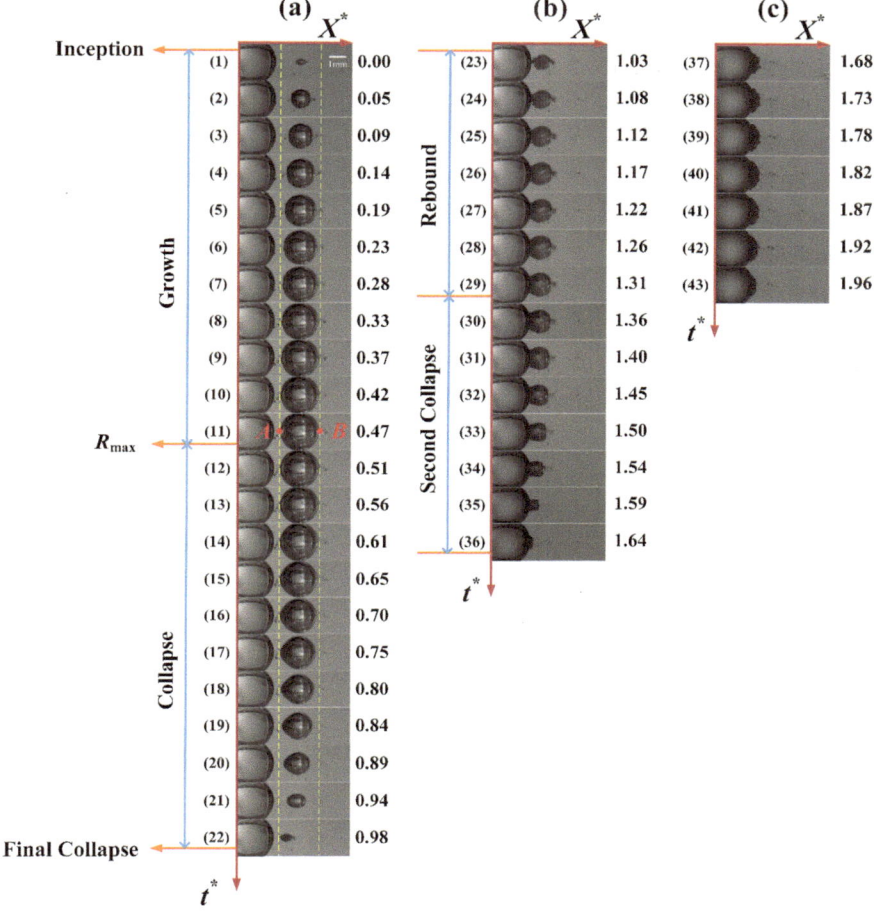

Fig. 5.5 The complete dynamics of a bubble near a particle. Reprinted with the permission from Ref. [2] Copyright (2023) (Springer Singapore)

collapse and splits into two parts at the end. When the bubble is further away from the particle, the bubble maintains a good spherical shape and the intensity of the collapsed jet is significantly reduced compared to the previous two cases.

Figure 5.7 classifies the morphology of the bubble near the particle in the first period based on the different bubble-particle distances. As shown in the figure, there are three typical collapse behaviors of the bubble which are mushroom-shaped, pear-shaped and spherical respectively.

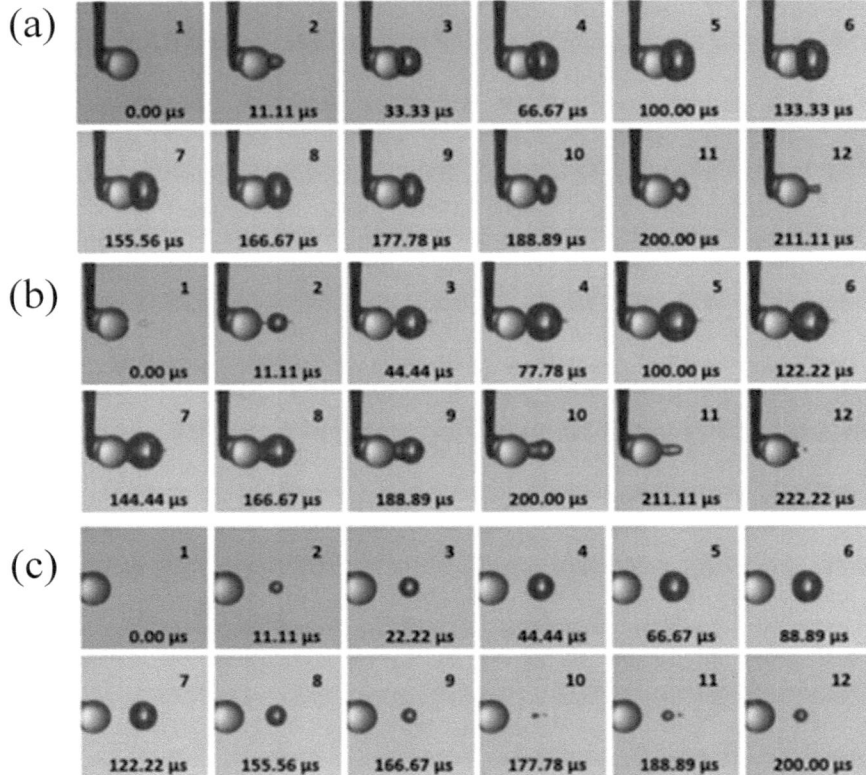

Fig. 5.6 The morphological evolution of the bubble during the first period. Reprinted with the permission from Ref. [3] Copyright (2018) (Elsevier)

5.2.3   Morphology Evolution in the Second Period

This section describes the morphological evolution of the bubbles during its second period, demonstrating the bubble dynamic behaviors such as bubble splitting, bouncing, and secondary collapse.

Figure 5.8 shows the morphological evolution of the bubble during the second period for different distances between the bubble and the particle. Figure 5.8a–c represent three cases from close to far from the particle, respectively. The first three frames of each subfigure show the initial, maximum and minimum moments of the first period of the bubble, and then the morphological evolution of the bubble in the second period is shown in detail. As shown in Fig. 5.8a, when the bubble is closer to the particle, the right side of the bubble wall forms a bulge outward in the second period, and when the bubble enters the rebound stage, the right side of the bubble is gradually rounded and grows to the maximum volume, finally occurs the second bubble collapse. As shown in Fig. 5.8b, when the bubble is at a medium distance

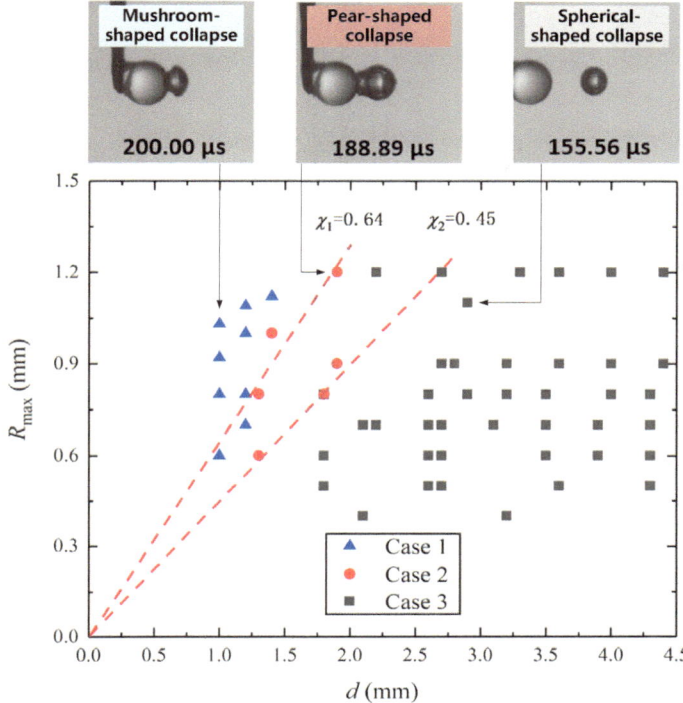

Fig. 5.7 Morphological classification of the bubble near the particle during collapse. Reprinted with the permission from Ref. [2] Copyright (2024) (Springer Singapore)

from the particle, the bubble in the rebound stage is spherical and accompanied by some irregular bumps on the left side, which gradually disappear as the rebound progresses. As shown in Fig. 5.8c, when the bubble is further away from the particle, the bubble splits into two parts with different sizes with the left side of the bubble is bigger. As the rebound process proceeds, the two bubbles gradually increase in size and move away from each other. During the second bubble collapse, the left side of the bubble is almost immobile while the right side of the bubble gradually moves to the right.

Figure 5.9 classifies the morphology of the bubble near the particle in the second period based on the bubble-particle distances. As shown in the figure, the dynamical behaviors of the bubble in the second period are classified into three typical cases, and give the parameter classification criteria for the three cases. When the size of the bubble is small, the ability to separate is weaker so that the boundary between cases 2 and 3 is not obvious.

Figure 5.10 shows the variation of the dimensionless second period time with the dimensionless distance between the particle and the bubble. As shown in the figure, the second period time decreases with increasing distance from the bubble to the particle in case 1 and case 2. For case 3, the second period of the left and the right

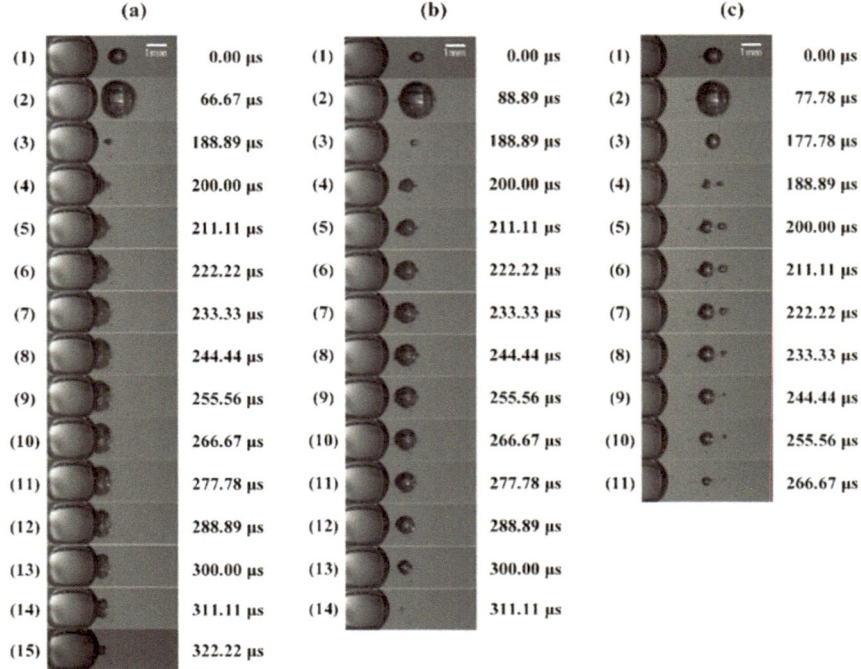

Fig. 5.8 Morphological evolution of the bubble during the second period. Reprinted with the permission from Ref. [4] Copyright (2023) (Springer Singapore)

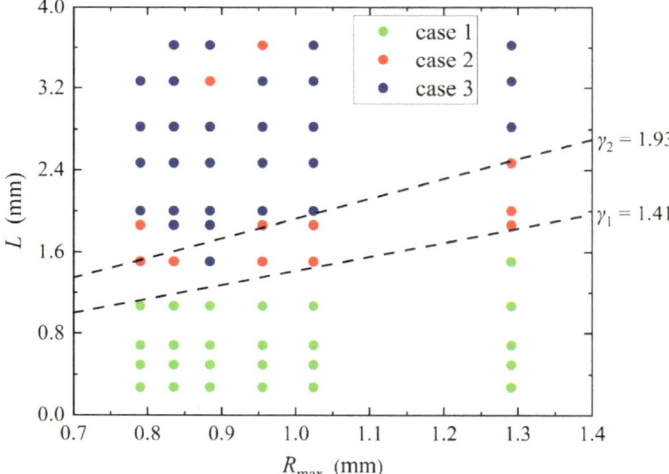

Fig. 5.9 The categorization of the dynamics of the cavitation bubble near the spherical particle into three cases. Reprinted with the permission from Ref. [4] Copyright (2023) (Springer Singapore)

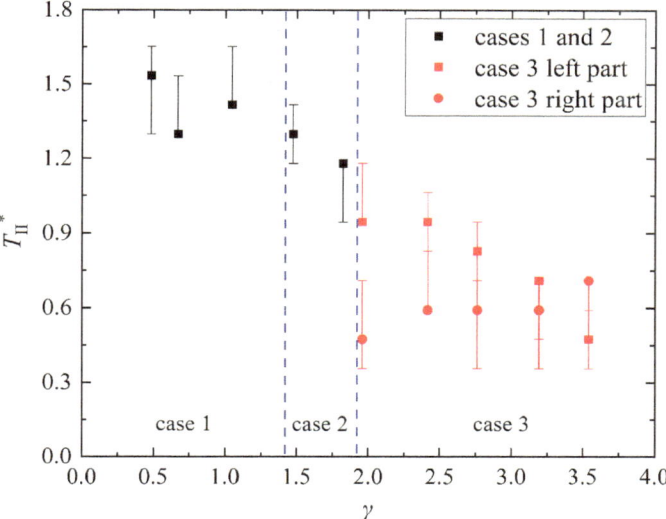

Fig. 5.10 The variation of the dimensionless second period time with the dimensionless distance between the particle and the bubble. Reprinted with the permission from Ref. [4] Copyright (2023) (Springer Singapore)

side of the bubble varies in the opposite way, which shortens and lengthens with increasing distance between the bubble and the particle, respectively. The second period time of the right bubble exceeds the left side bubble when the bubble is farther away from the particle.

5.3 Jet

5.3.1 Forward Jet

Figure 5.11 shows the evolution of the morphology of the bubble for different distances between the bubble and the bulge. As shown in Fig. 5.11a–c, the bubble–bulge distance has a significant effect on the jet. When the bubble is close to the bulge (Fig. 5.11a), the non-spherical deformation and movement of the bubble in the collapse phase is significant and the direction of the jet is biased to the upper left. As the distance between the bubble and the bulge increases, the degree of non-spherical deformation and movement of the bubble gradually decreases, and the direction of the jet is gradually perpendicular to the flat wall.

Figure 5.12 shows the variation in the direction of bubble movement. As shown in the figure, the larger the azimuthal angle of the bubble relative to the bulge, the direction of bubble movement first increases with it and reaches a peak, and then

(a) $l^* - 3.22$

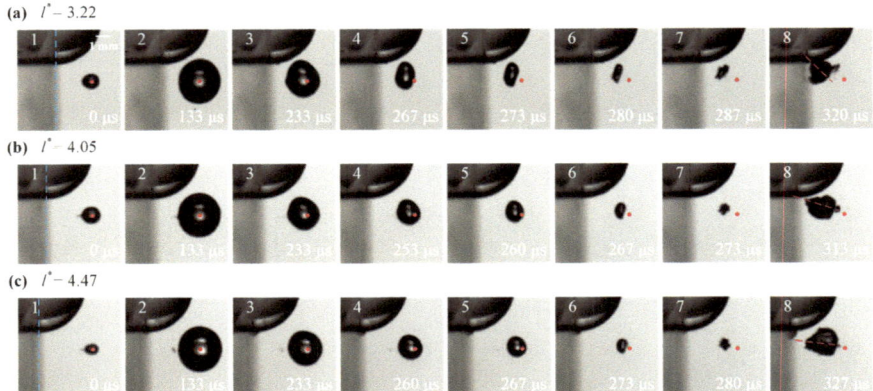

(b) $l^* - 4.05$

(c) $l^* - 4.47$

Fig. 5.11 The evolution of the morphology of the bubble for different distances between the bubble and the particle. Reprinted with the permission from Ref. [5] Copyright (2024) (American Institute of Physics)

decreases to zero. In addition, as the distance between the bubble and the bulge increases, the influence of the bulge on the bubble diminishes and the change in the direction of bubble movement increases.

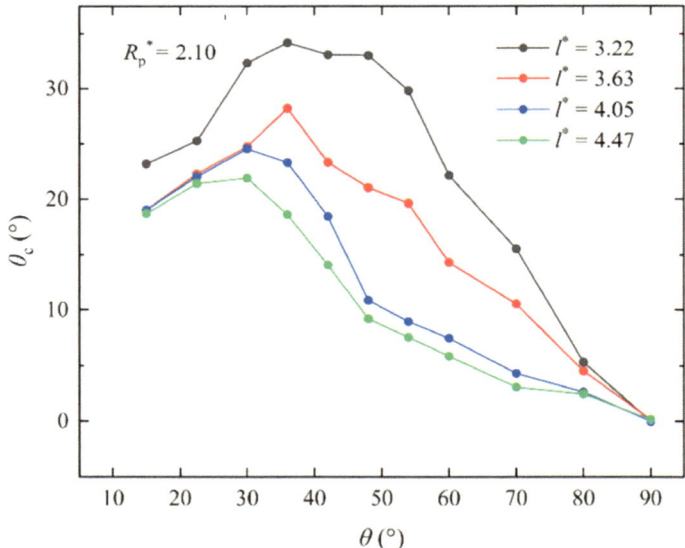

Fig. 5.12 The variation in the direction of bubble movement. Reprinted with the permission from Ref. [5] Copyright (2024) (American Institute of Physics)

5.3.2 *Counter Jet*

Figure 5.13 shows the dynamics of the counter jet during the second period for a bubble near the particle. As shown in Fig. 5.13a, the generation of the counter jet is often observed when the bubble is close to the particle surface. The process of generation can be explained by the interaction between the shock wave and the bubble. The bubble is pierced when it contacts the particle surface. And, at the end of the bubble collapse, an annular bubble is formed with a shock wave generated. The shock wave generates high pressure at the point of encounter, which in turn triggers a tension wave that leads to a counter jet. In addition, the curvature of the particle surface affects the reflection and interaction of the shock wave, which is different from the case of a flat boundary. Thus, there is a significant difference between the counter jet (generated by the bubble near the particle) and the faward jet.

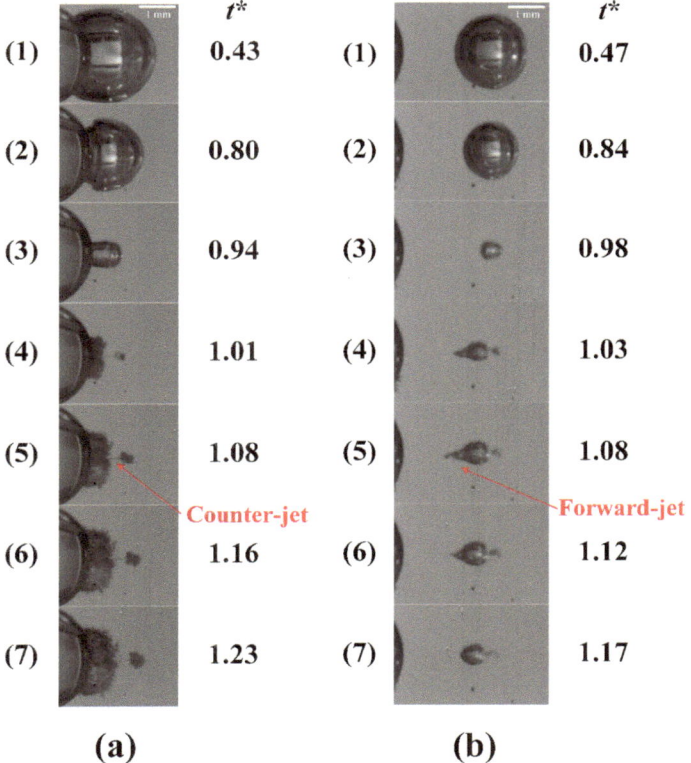

Fig. 5.13 The dynamics of the counter jet during the second period for a bubble near the particle. Reprinted with the permission from Ref. [4] Copyright (2023) (Springer Singapore)

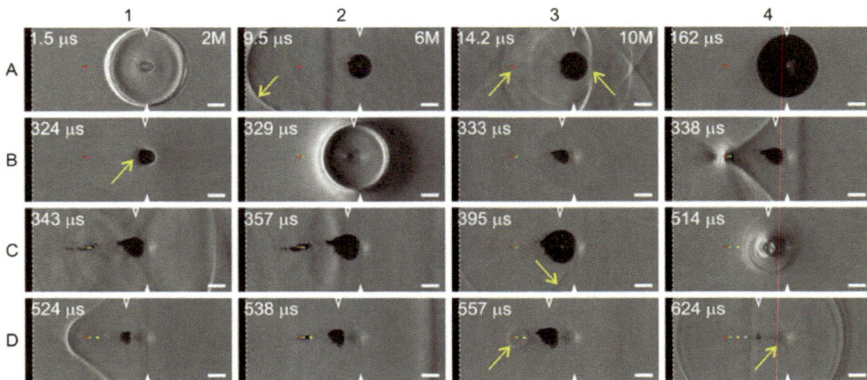

Fig. 5.14 The evolution of a shock wave near a recessed wall surface. Reprinted with the permission from Ref. [7] Copyright (2021) (Elsevier)

5.4 Shock Wave

5.4.1 Shock Waves During Bubble Inception

Geng et al. [6] investigated the process of bubble and shock wave generation, development and expansion in a free liquid. The experimental results show that at the early stage of bubble expansion, both the bubble and the shock wave develop with a non-spherically symmetric law. In the subsequent evolution, both the bubble and the shock wave gradually become spherically symmetric. In addition, the laser energy has a significant effect on the plasma length. When the energy is higher, the cone zone is longer. In addition, Požar et al. [7] found that after reflection from the wall, the shock wave may have a significant impact on the collapse behaviors of the bubble. Figure 5.14 shows the evolution of a shock wave near a depression wall. The shock wave generated by the bubble inception is reflected by the depression wall and converges near the bubble. The reflected shock wave interacts with the bubble and causes the bubble to collapse faster.

5.4.2 Shock Waves During Bubble Collapse

Figure 5.15 shows the propagation of the collapse shock wave for different distances between the bubble and the wall. Figure 5.15a–c correspond to the case where the bubble is further and further away from the wall. As shown in Fig. 5.15a, there are two obvious spherical waves and other weaker spherical fronts, and it can be observed that the two stronger spherical waves are formed by the superposition of many weaker shock waves. In addition, these two spherical waves overlap at the right side. In

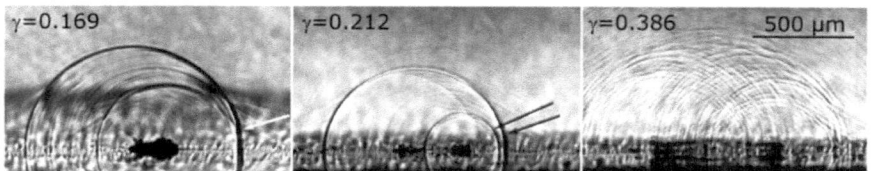

Fig. 5.15 The shock waves in the bubble collapse process. Reprinted with the permission from Ref. [8] Copyright (2022) (Elsevier)

Fig. 5.15b, the distance of the bubble from the wall increases, the intensity of these two more obvious spherical waves decreases and the degree of overlap decreases on the right side. In Fig. 5.15c, there are no longer two distinct shock waves but a series of individual shock waves.

Figure 5.16 shows the shock waves and the pressure characteristics of the bubble collapse process for different bubble-particle distances. Figure 5.16a represents the evolution of the shock wave near the wall without particles. At this point, the bubble is similar with the collapsing in the free-flow field, forming a spherical shock wave outwards generating the pressure peak as shown in the figure. In Fig. 5.16b, a neck-like structure is formed and us further broken, he subvacuole near the particle collapses faster to form a shock wave. The annular jet forms later and hits the wall, forming a water hammer shock wave. In Fig. 5.16c, the distance of the bubble from the particle is further increased, and the side closer to the particle collapses faster, forming a spherical monolayer shock wave and bringing about the first pressure peak. Then, the second pressure peak corresponding to the collapse of the other surface of the bubble will be demonstrated. Compared with Fig. 5.16c, the shape of the collapse of the left side of the bubble contacting the particle surface in Fig. 5.16d becomes flatter and more elongated, creating a layered shock wave. The first pressure peak is also caused by the collapse of the bubble close to the particles and two peaks appear subsequently due to the layered shock wave. In Fig. 5.16e, when the bubble collapses to its minimum volume and separates from the particle, a clear layer separation effect occurs and the pressure peak increases significantly. In Fig. 5.16f, the layer effect is weakened, the two layers of collapsed shock waves are closer, and the secondary pressure peak almost disappears. In Fig. 5.16g–i, the bubble forms a single-layer spherical shockwave during collapse, corresponding to one main pressure peak.

5.5 Droplet Splash

Figure 5.17 shows the collapse of a cavitation bubble within an attached droplet and its induced droplet splashes. In Fig. 5.17, the maximum radius of the bubble is larger compared to the droplet radius. As shown in Fig. 5.17, the secondary nucleation phenomenon is obvious at the early stage of bubble growth (frames 1–3). In addition, as the bubble grows (frames 4–7), the stability of the top surface of the droplet is

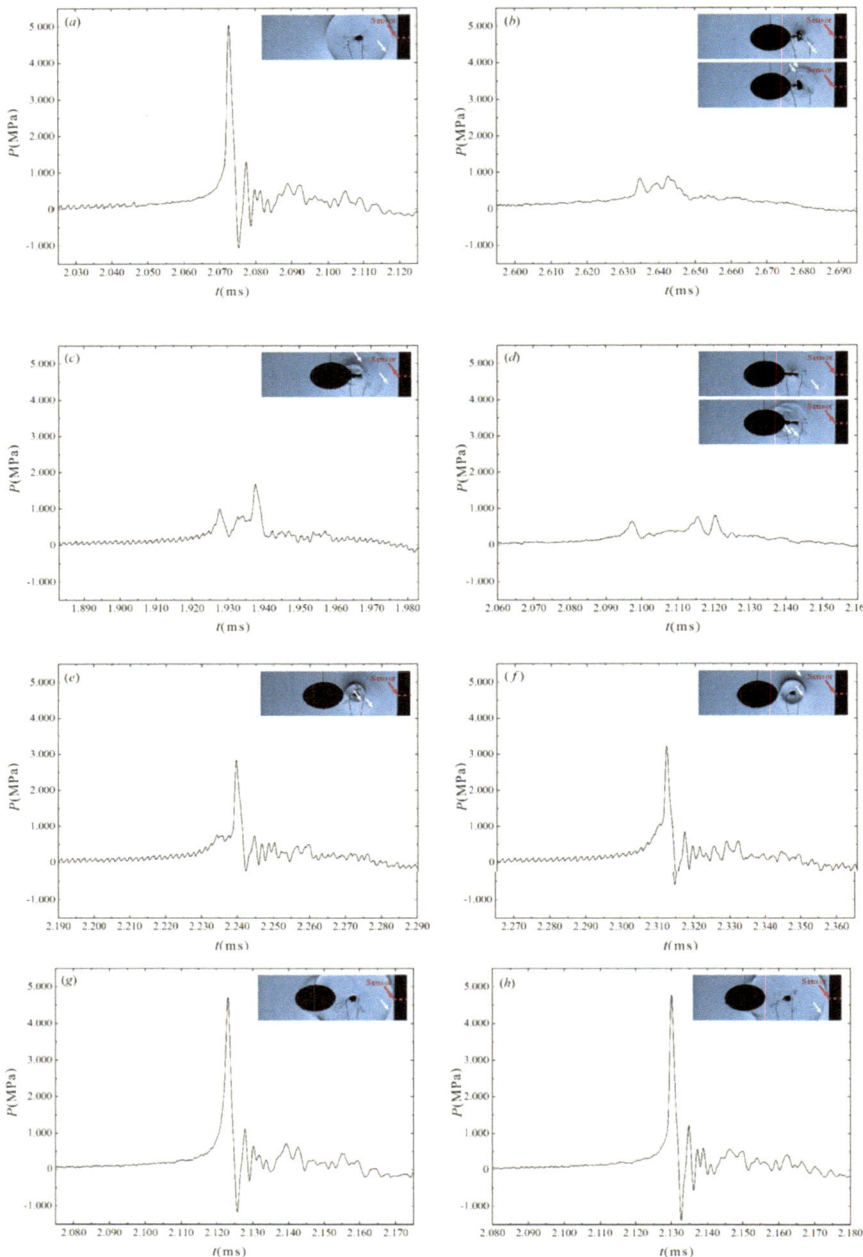

Fig. 5.16 The shock waves and the pressure characteristics of the bubble collapse process for different bubble-particle distances. Reprinted with the permission from Ref. [9] Copyright (2023) (Elsevier)

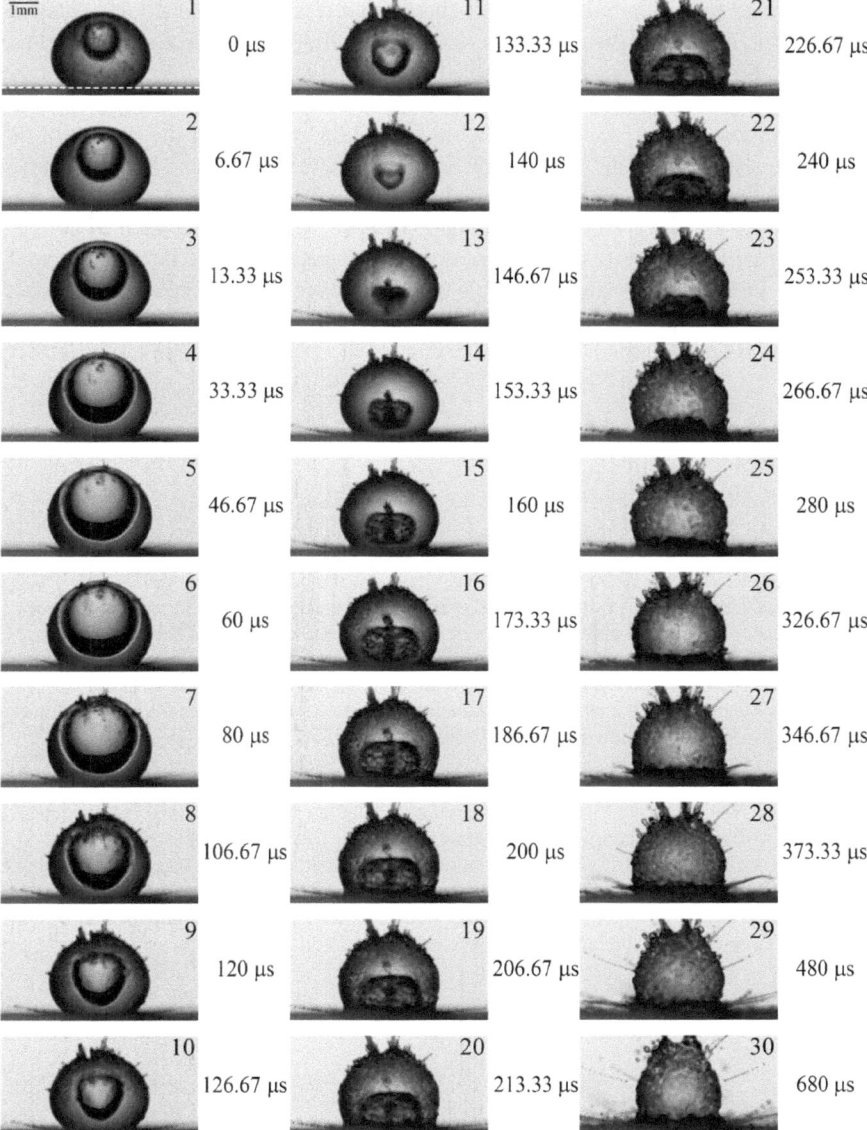

Fig. 5.17 Collapse of a cavitation bubble within an attached droplet and its induced droplet splashes

destroyed due to the intense growth of the bubble (frame 6), and "skirt" splashes appear at the bottom of the droplet in contact with the hydrophobic surface (frame 7). During the first collapse of the bubble (frames 8–12), a depression appears at the top of the bubble and moves towards the wall. In addition, multiple splashes begin to appear on the top surface of the droplet, and "skirt" splashes at the bottom of the droplet are also developing. Subsequently, during the rebound of the bubble (frames 13–20), a jet towards the wall appears and the bubble is penetrated. A reverse jet will also appear (frame 13). During the second collapse of the bubble (frames 21–27), the bubble will continue to move toward the wall and eventually contact the wall. Due to the violent collapse of the bubble towards the wall, "skirt" splashes are again induced. In addition, the instability of the droplet surface gradually increases, and multiple splashes are added to the surface and the morphology changes significantly.

References

1. Wang X, Li S, Shen J et al (2024) Dynamic behaviors of a bubble near a rectangular wall with a bulge. Phys Fluids 36(2):1241
2. Yu J, Wang X, Shen J et al (2024) Physics of cavitation near particles. J Hydrodyn 36(1):102–118
3. Zhang Y, Chen F, Zhang Y et al (2018) Experimental investigations of interactions between a laser-induced cavitation bubble and a spherical particle. Exp Therm Fluid Sci 98:645–661
4. Wang X, Su H, Li S et al (2023) Experimental research of the cavitation bubble dynamics during the second oscillation period near a spherical particle. J Hydrodyn 35(4):700–711
5. Wang X, Zhang C, Shen J et al (2024) Influence of a hemispherical bulge on a flat wall upon the collapse jet of cavitation bubbles. Phys Fluids 36(3):323
6. Geng S, Yao Z, Zhong Q et al (2021) Propagation of shock wave at the cavitation bubble expansion stage induced by a nanosecond laser pulse. J Fluids Eng 143(5):051209
7. Požar T, Agrež V (2021) Laser-induced cavitation bubbles and shock waves in water near a concave surface. Ultrason Sonochem 73:105456
8. Reuter F, Deiter C, Ohl CD (2022) Cavitation erosion by shockwave self-focusing of a single bubble. Ultrason Sonochem 90:106131
9. Zou L, Luo J, Xu W et al (2023) Experimental study on influence of particle shape on shockwave from collapse of cavitation bubble. Ultrason Sonochem 101:106693

Chapter 6
Conclusion

This book comprehensively reviews the oscillation and collapse properties of the bubble under different conditions, as well as the physical mechanisms involved in the process based on the experimental, theoretical and numerical methods. Through in-depth analyses of the whole book, this book reveals the complexity and diversity of the behavior of the bubble and the following main conclusions are obtained.

Firstly, the radial equations of motion of spherical and cylindrical bubbles as well as the bubble in a liquid droplet are derived from different perspectives. The effects of parameters such as fluid compressibility and viscosity are taken into account to establish mathematical models describing the motion of different kinds of bubbles in fluids. This provides mathematical methods for understanding the dynamic behavior of various bubbles.

Secondly, the resonant properties of a cavitation bubble oscillating in an acoustic field are investigated based on the perturbation method, multi-scale method and Laplace transform method. There is a significant difference in the intrinsic frequency and damping coefficient between the spherical and interstitial bubble in the weakly nonlinear oscillations. The study of strong nonlinear oscillations shows that the peak values of bubble radius and velocity increase and then decrease, and the rate of change with time accelerates. Under single-frequency excitation, the unstable region of the frequency response curve expands with increasing amplitude.

Thirdly, the Kelvin impulse theory that can be applied to predict the behaviour of bubble collapse is presented. The basic assumptions, derivation process and formulas of the Kelvin impulse theory model are presented. Subsequently, boundary treatment methods, such as the Weiss theorem, the image method and the conformal transformation, are given. Furthermore, several typical Kelvin impulse results for a bubble under different boundary conditions (such as a bubble near a particle, near a bulge wall, within a droplet and near a hydrofoil), are described. It is found that the collapse characteristics of the bubble are significantly affected by the boundary conditions, and the direction and intensity of the Kelvin impulse are closely related to the relative

© The Author(s), under exclusive license to Springer Nature Switzerland AG 2024   93
X. Wang et al., *Fundamentals of Single Cavitation Bubble Dynamics*,
SpringerBriefs in Energy, https://doi.org/10.1007/978-3-031-75041-0_6

positions of the bubble and the boundary, as well as the curvature, shape and other factors.

Finally, the morphological evolution, jet and shock wave characteristics of a cavitation bubble are investigated by high-speed photography experiments. The composition of the high-speed photography platform and the layout of the experimental platform are described in detail. The dynamic changes and typical evolution characteristics of the bubble during the first and second periods of oscillation are analyzed and compared. The characteristics of the forward and reverse jets and their evolutions are revealed. Then, the formation mechanisms and propagation behavior of the shock waves generated by the bubble at different stages are investigated. In addition, it is observed that the bubble collapse process in a droplet is accompanied by the complex droplet splashing phenomenon.

Index

© The Editor(s) (if applicable) and The Author(s), under exclusive license
to Springer Nature Switzerland AG 2024
X. Wang et al., *Fundamentals of Single Cavitation Bubble Dynamics*,
SpringerBriefs in Energy, https://doi.org/10.1007/978-3-031-75041-0